THE ORIGIN, EXPANSION, AND DEMISE OF PLANT SPECIES

Donald A. Levin

New York   Oxford

OXFORD UNIVERSITY PRESS

2000

Oxford University Press

Oxford New York
Athens Auckland Bangkok Bogotá Buenos Aires Calcutta
Cape Town Chennai Dar es Salaam Delhi Florence Hong Kong Istanbul
Karachi Kuala Lumpur Madrid Melbourne Mexico City Mumbai
Nairobi Paris São Paulo Singapore Taipei Tokyo Toronto Warsaw

and associated companies in
Berlin Ibadan

Copyright © 2000 by Oxford University Press

Published by Oxford University Press, Inc.
198 Madison Avenue, New York, New York 10016

Oxford is a registered trademark of Oxford University Press

Library of Congress Cataloging-in-Publication Data
Levin, Donald A.
    The origin, expansion, and demise of
plant species / Donald A. Levin.
        p. cm. – (Oxford series in ecology and evolution)
    ISBN 0-19-512728-5; ISBN 0-19-512729-3 (pbk.)
    1. Plants–Evolution. 2. Plant species. 3. Plants–Extinction.
I. Title. II. Series.
    QK980 .L48 2000
    581.3'8–dc21   99-34861

1 2 3 4 5 6 7 8 9
Printed in the United States of America
on acid-free, recycled paper

Preface

This book is an exploration of species as dynamic entities. Specifically, it is an exploration of the species' passage from birth to death. Each species has its own unique passage that is affected by its gene pool, dispersal ability, interactions with competitors and pests, and the habitats and climatic conditions to which it is exposed.

We may identify stages in the lives of species. The first is the birth or origin of species. Each species has a beginning in a geographically defined context. The second stage is expansion. All successful species spread from their focus of origin. The third stage is differentiation and loss of cohesion. Most species undergo some form of geographical differentiation, even if the most divergent entities are not formally recognized. In the process, the ability of populations to exchange genes is reduced. The fourth stage is decline and extinction. All species eventually come to an end following the contraction and fragmentation of a once larger range. This book is arranged so that it follows these stages.

Chapter 1 provides a broad rationale for the aforementioned approach and includes a discussion of species concepts. Chapter 2 focuses on the ecological transition, which is an essential part of the speciation process. A highlight of the chapter is the evolutionary lability of reproductive and vegetative characters. Patterns of radiation also are discussed. Chapter 3 is concerned with the genetic transition that accompanies speciation. The prime issues are the genetic basis for species differences in ecological and reproductive attributes, the genetic and chromosomal bases for postpollination reproductive barriers, and the rate of genetic transitions. Chapter 4 concerns the geographical scale of speciation. The primary issue is whether speciation is a local or regional phenomenon. I argue for the former. Chapter 5 deals with the geographical expansion of neospecies. In addition to describing neospecies per se, migrations of introduced species and postglacial migrations are discussed as these entities are useful neospecies surrogates. Chapter 6 is concerned with the differentiation and breakdown of species unity. Focal points are patterns of differentiation, the origin of barriers to gene exchange, and the consequences of contact between population systems. Chapter 7 deals with the decline and demise of species. Considerable attention is given to the ecology and genetics of rare species, and factors contributing to species extinction. Chapter 8 focuses on the fate of incipient species, most of which are likely to go extinct before they develop into full-fledged species. The role of the progenitor in their extinction is the prime issue here. Chapter 9 deals with macroevolutionary issues such as speciation rates, extinction rates, and species durations.

I am most grateful for the assistance given to me during the formulation of this book. I wish to give special thanks to Norman Ellstrand, Gábor Lendvai, and Uzi Plitmann who read the entire manuscript (often more than one version) and provided many very insightful and useful suggestions. Many thanks to Beryl Simpson and Dan Crawford for commenting on several chapters. Thanks also to the series editors, Paul Harvey and Bob May, for their considerable efforts in my behalf. I bear full responsibility for the omissions and errors in the final product.

I am also most grateful to the people who helped in the production of this book. Many thanks to the Department of Botany secretaries Suzanne Seiders and Kathleen Fordyce and to the department artist Marianna Grenadier, who redrew all of the figures used in the book from the original sources. Also a large thanks to Bob Nagy, who kept my computer functioning during the term of the project.

Thanks as well to the copyright holders for their permission to redraw figures.

Finally a special thanks to M. Marjoree Menefee.

I dedicate this book to my wife, Georgia, and my children, Phil and Debra.

Austin, Texas   D.A.L.
April, 1999

Contents

The Origin, Expansion, and Demise of Plant Species

1

The Premise and Species Concepts

Organic diversity is divided into nodes or clusters that, for the most part, are discontinuous. These clusters are assigned to categories such as races, subspecies, species, genera, and families, which in turn are placed in a phylogenetic framework. The most important category is that of the species, which is the basic unit in evolutionary biology. For decades species have been thought of as the "keystone of evolution," "the entities which specialize, which become adapted or which shift their adaptation" (Mayr, 1969). The species has had a special status as an evolutionary unit because it is thought to be more natural than higher categories and more amenable to definition and empirical demonstration (Huxley, 1940; Dobzhansky, 1940; Mayr, 1942). Moreover, the species is considered to have properties of its own—that is, not shared with its component individuals—such as breeding systems, mate recognition systems, and distributions (Grant, 1981; Paterson, 1985). Species also are special because they provide the only mechanism for protecting (by isolating gene pools from other gene pools) phenotypic and underlying genetic change that accumulate by selection and genetic drift (Futuyma, 1987).

The species has been and remains the subject of heated debate in terms of definition and origin. This is exemplified in books such as *Animal Species and Evolution* (Mayr, 1963), *Plant Speciation* (Grant, 1981), *Mechanisms of Speciation* (Barigozzi, 1982), *Speciation and its Consequences* (Otte and Endler, 1989), *Endless Forms* (Howard and Berlocher, 1998), and in symposium publications such as one on species concepts and phylogenetic analysis in *Systematic Botany* (1995).

Regardless of how you define them or their mode of origin, species persist in time. The fate of species in time has been considered largely by paleontologists. They have made three general observations important to the issues developed in this book. First, in some groups, species appear to undergo little morphological change over long time spans, whereas in other groups species have undergone gradual (progressive) change in time (Stanley, 1979; Eldredge, 1989; Vrba, 1980). Second, the location and sizes of species ranges may change considerably in time. Third, species with certain traits may persist longer than species with other traits; and species with certain traits may spawn more new species than species with other traits.

Microevolutionists have written little about the fate of species, or more precisely the evolutionary history of species from their inception to their death. As a result we lack a general understanding of species as dynamic evolutionary units beginning somewhere with a set of traits, having continuity in time, and eventually going extinct. The focal point of almost all studies has been a process or pattern or interaction that is explained by proximal evolutionary and environmental events that need not take one very far back in the evolutionary history of the species. When we have a paleontological history of species showing an approximate time of birth and death, and showing changes in morphological variation, abundance, and range, we cannot directly associate them with processes.

The Premise

The evolutionary histories of species have general aspects in common, even though the species may differ greatly in their attributes. Each species has a beginning in some geographically defined context: a population, or an aggregation of populations. This geographical entity has a location. All successful species spread from their focus of origin. Most species undergo some geographical differentiation, even if the most divergent entities are not recognized as different races or subspecies. All species interact with other species. All species eventually come to an end following the contraction and/or fragmentation of a once larger range.

Each species has its own unique passage from birth to maturity to death. The nature of this passage will be affected by the character and amount of genetic variation it begins with and accumulates; its developmental flexibility; its dispersal abilities; its interactions with symbionts, competitors, related species, and with pathogens, parasites and predators, the habitats and climatic conditions it encounters in time and space, and its ability to track (adjust to) environmental change.

The idiosyncracy of the passage from birth to death for species also has its origins in their evolution. There are two contributory factors. First, the sequence of genetical and demographic changes that preceded the "birth" of a new species (neospecies) differs substantially among species. Second, the genetical properties of the progenitor vary among species, even if we are talking about sister species. Given that neospecies are variations on parental themes and that the parental species differ, neospecies must differ in their potentialities.

We have very little direct information on the evolutionary histories of species from their inception. We are restricted to species that evolved in recent times. Nevertheless, we may identify stages in the lives (or histories) of all species based on our understanding of different species as we see them today. These stages are (1) birth or origin, (2) expansion, (3) differentiation and the loss of cohesion, and (4) decline and extinction. The first two stages of species' lives are perhaps the easiest to understand using the geographical and ecological relationships of neospecies and progenitors. Given that newly emergent species tend to be rare, we can gain valuable insights about their attributes and limitations from investigations of other rare species. This is not to suggest that all rare species are of recent origin, which of course they are not. The expansions of newly emergent species from narrow geographical centers may be understood in part from patterns of invasions following introduction of weedy and nonweedy species into hospitable areas.

The extent to which expanding species undergo geographical differentiation depends on species' gene pools, breeding systems, dispersal patterns, demographic properties, and on ecological opportunity. Certain combinations of features are conducive to adaptive differentiation, others are conducive to nonadaptive differentiation, and others mediate against differentiation. I will discuss these combinations and examples of species illustrative of them.

Populations of newly emerged species contain considerable similarity, although they may be separated in space. As neospecies age, the forces that integrate species tend to become weaker and groups of populations become more genetically independent. The distribution of populations also may become less continuous over time, thus isolating groups of populations by distance.

The aging of species often is manifested in the contraction and fragmentation of species' ranges. There are many species that seem to be in decline and that are being studied from ecological and genetical perspectives. Some are threatened by unrelated competitors, some by pathogens and predators, and some even by congeneric species. A paucity of genetic variation or inbreeding depression may contribute to species decline. Recent extinction has been documented for a few species, although specific causation is difficult to attribute.

The underlying premise of this book is that a species is a dynamic entity that undergoes alterations in its gene pool, variation pattern, and geographical distribution, and perhaps also undergoes alterations in its habitat preferences, and in its relationships with mutualists and pests. The evolutionary history of a species is an outcome of the interaction of the species' biological attributes with an environment that changes in space and time. As such, this history will not be understood if species are treated only as taxonomic units, evolutionary units, or ecological units. Thus it is best to take a pluralist approach to species' passages in time, combining genetical and ecological perspectives.

In attempting to describe the dynamics of species' histories, this book is arranged so that it follows the stages in the lives of species from their inception to their demise. All species, regardless of their positions in the plant and animal kingdoms, will proceed through these stages, unless they go extinct early in their existence. This book will deal with this passage within the context of flowering plants.

The fate of a population system we call a species in part depends on our species concept. One concept may yield a very different result from another. Accordingly, the remainder of this chapter will deal with species concepts. First,

I will review the concepts most often employed by evolutionists, then I will propose some alternative viewpoints.

Species Concepts

Species concepts can be classified as mechanistic, historical, or phenetic. Mechanistic concepts are based on speciation processes. They also contain the idea of cohesion and the mechanisms that "hold species together." Historical species concepts diagnose species on the basis of character differences and ancestry. Phenetic species concepts emphasize the evidence employed to recognize species. They are independent of theories concerning the origin and maintenance of species. All species concepts, explicitly or implicitly, include the notion of idiosyncracy and independence from other species.

A Synopsis of Species Concepts

Mechanistic Concepts

The best known and most influential mechanistic concept is Mayr's (1963) Biological Species Concept. He defined the biological species as "groups of actually or potentially interbreeding natural populations which are reproductively isolated from other such groups." Nevertheless he recognized, indeed argued, that most species have distinct ecological niches. The "evolutionary significance of species" lies in their ecological uniqueness (Mayr, 1970). Mayr (1982a) later amended the BSC to include the idea that species also would have divergent ecological requirements (tolerances). He wrote that "a species is a reproductive community of populations (reproductively isolated from others) that occupies a specific niche in nature."

The idea that species are reproductive communities was extended by Paterson (1985) in his Recognition Species Concept. He states that a species is "the most inclusive population of individual biparental organisms which share a common fertilization system." Different members of the same species recognize each other as potential mates. Speciation involves the alteration of a specific-mate recognition system—that is, the signal-response interactions that serve to bring sexual partners together for mating. The specific-mate recognition system is a subsystem of the fertilization system. Species are not defined in terms of isolation. Paterson sees reproductive isolation as an incidental by-product of divergence in the mate recognition systems of populations. Paterson (1985, 1986) contends that signaling/response interactions are most likely to change when animal species encounter new habitats. Thus species are likely to differ in both reproductive and ecological attributes.

Templeton (1989) refined the union of reproductive and ecological criteria in his Cohesion Species Concept. According to the concept, a species is "the most inclusive population of individuals having the potential for cohesion through intrinsic cohesion mechanisms." Members of a species are integrated by genetic and demographic exchangeability. The degree of each may vary among species. Members of a species share a fertilization system in the sense of Paterson (1985) and the same intrinsic ecological tolerances or fundamental niche (Hutchinson, 1965). Genetic exchangeability is promoted by sexual reproduction, especially outcrossing, and by the absence of isolating mechanisms. It is required for the spread of novel genetic variants by gene flow. Demographic exchangeability is promoted by factors constraining the ability of populations to respond to alternative selective pressures and genetic drift. Demographic exchangeability is required for natural selection and genetic drift to operate. Species retain their integrity through some combination of ecological and genetic barriers. Different species lack demographic and genetic exchangability.

The importance of ecological differences in species delineation was emphasized by Van Valen (1976). He proposed an Ecological Species Concept in which a species is a "lineage (or closely related set of lineages) which occupies an adaptive zone minimally different from that of any other lineage in its range, and which evolves separately from all other lineages outside its range." An adaptive zone is considered to be some part of the resource space plus parasites and predators. According to the ESC, speciation involves the invasion of new niches or the division of existing niches. The phenetic clusters recognized as species owe their discreteness to evolutionary responses to discontinuities in the environment. The utility of an ecological species concept has been discussed by other authors (Raven, 1978, 1980; Andersson, 1990). The concept is best applied to apomictic population systems where sexual reproduction does not occur, and thus where genomic discord cannot be manifested.

Species defined by one mechanistic criterion may not correlate well with those defined by other such criteria. Moreover, species defined by mechanistic criteria may not correlate well with those defined by historical or phenetic criteria.

Historical Concepts

The two primary historical species concepts are the Evolutionary Species Concept and the Phenetic Species Concept. The Evolutionary Species as defined by Simpson (1961) and Wiley (1981) is an independent lineage that has its own "unitary evolutionary role" (ecological tolerance) and evolutionary tendency. According to the most recent Phylogenetic Species Concept (Nixon and Wheeler, 1990), a species is "the smallest aggregation of populations (sexual) or lineages (asexual) diagnosable by a unique combination of character states in comparable individuals (semaphoronts)." Earlier phylogenetic species concepts have been discussed with reference to plants by Rieseberg and Brouillet (1994), Baum and Donoghue (1995), Luckow (1995), McDade (1995), and Olmstead (1995).

Baum and Shaw (1995) formulated the most recent historical concept, the Genealogical Species Concept. They propose that "species [are] basal, exclusive groups of organisms" in which "[a] group of organisms is exclusive if their genes coalesce [can be traced to a common ancestral gene] more recently within the group than between any member of the group and any organisms outside the group."

Phenetic Species Concepts

As defined by Michener (1970), "A species is a group of organisms not itself divisible by phenetic gaps resulting from discordant differences in character states (except for morphs such as those resulting from sex, caste, or age differences), but separated by such phenetic gaps from other such groups." This concept was updated by Mallet (1995) to include genetic data. He proposed that "species . . . are . . . identifiable genotypic clusters . . . recognized by a deficit of intermediates, both at single loci (heterozygote deficits) and at multiple loci (strong correlations or disequilibria between loci that are divergent between clusters)."

Whereas species concepts can be grouped as mechanistic, historical, or phenetic, they also can be grouped as monistic and pluralistic, as discussed by Shaw (1998). Monistic species concepts allow that there is one kind of species entity with many biological properties that arise in a multitude of different ways. Pluralistic concepts allow that there are several kinds of species, each arising from the same evolutionary process. The Evolutionary Species Concept and Cohesion Species Concept are examples of monistic concepts; and the Biological Species Concept and the Recognition Species Concept are examples of pluralist concepts. Shaw advocates pluralistic concepts because single-criterion entities are readily definable and attributable to one evolutionary process. The problem, of course, is to decide which single criterion should be applied.

Weaknesses of Species Concepts

Mechanistic species concepts are more amenable to the study of speciation than historical species concepts. The former involves the action of evolutionary forces on gene pools, whereas the latter deal with the manifestations of reproductive and ecological cohesion, and cohesion due to common ancestry, rather than the evolutionary processes responsible for cohesion (Templeton, 1989). Thus mechanistic concepts are more suitable for the present discussion of species evolution and will be considered below to the exclusion of others.

An appropriate species concept must be compatible with processes that allow new species to be successful independent entities. The aforementioned mechanistic concepts fail to do so in different respects. The original Biological Species Concept is found wanting because it defines a species solely in terms of its reproductive relationship with another species. Species by definition are reproductively isolated. The Recognition Species Concept is found wanting because it basically involves changes in the signaling/response interactions between pollen and pistil (Paterson, 1993). Paterson (1985, 1986) contends that signaling/response interactions may change when animal species encounter new habitats. However, there is no reason why pollen-pistil compatibilities or specificities would change when plant species invaded new habitats, unless these genetically based dictates were affected by stochastic processes (de Nettancourt, 1977; Levin, 1978a).

The Ecological Species Concept is found wanting because a shift in ecological tolerance alone would not preclude large-scale interbreeding between juxtaposed populations. Although migrants would be at a selective disadvantage, there could be considerable exchange of genes for traits unrelated to the environmental difference, thus violating the notion of species as independent gene pools. Surely one would not argue that the evolution of heavy-metal tolerant populations from "normal" pasture populations is tantamount to speciation (Bradshaw and McNeilly, 1981).

The Cohesion Species Concept (Templeton, 1989) is appealing because it embraces both pivotal aspects of species' biology, reproduction, and resource acquisition. However, all species are not equal in the sense that some may be integrated primarily by ecological factors, whereas others may be integrated primarily by genetic factors.

Hybridization, autopolyploidy, and chromosomal rearrangements, and asexual reproduction are problematic for most species concepts. The Biological Species Concept, Cohesion Concept, and Recognition Concept would not recognize species that engaged in extensive hybridization as good species. Hybridization would not deny species status to Ecological Species Concept unless it threatened the integrity of the species.

If a population system differed from its progenitor by virtue of autopolyploidy or novel chromosomal rearrangements, species status would be accorded by the Biological Species and Cohesion Species Concepts, but not by the Ecological Species Concept. If chromosomal change did not foster cross-incompatibility, then chromosomally distinct populations would not be recognized as different species according to the Recognition Concept. Autopolyploidy probably would introduce partial cross-incompatibility, chromosomal rearrangements probably would not.

Distinctive asexually reproducing entities that had well-differentiated ecological profiles would be given species status under the Ecological and Cohesion Concepts. These entities could not be classified using the criteria of the Biological or Recognition, Concepts. An alternate concept would have to be employed for such plants.

The Practical Reality of Species

It is not surprising that no species concept is suitable for the contingencies of plant diversity. Whereas the processes of evolution are universal, the products are highly idiosyncratic owing to their different ancestries, potentialities, and evolutionary histories. The historical relationships and ecological and genomic discontinuities of population systems are nearly as diverse as the entities one is attempting to classify. Thus species in one lineage may be quite different from species in another, as may be species within the same lineage.

We must keep in mind that species concepts are just concepts, powerful as they may be in organizing diversity into distinctive packages. Any attempt to neatly fit biological diversity into any single species concept is likely to be futile.

Ehrlich and Raven (1969) concluded that "the idea of good species . . . is a generality without foundation—an artifact of the procedures of taxonomy. These procedures require that distinct clusters be found and assigned to some level in a hierarchy—subspecies, species . . . and so on." In turn the taxonomic system communicates little about the organism being considered, although it appears to communicate a great deal. Raven, Berlin, and Breedlove (1971) contend that "our systems of names appears to achieve a reality which it does not in fact possess."

As with all theoretical concepts, species concepts bear within themselves the character of instruments (Levin, 1979). In the final analysis they are only tools that are fashioned for characterizing organic diversity. Concentrating on the tools draws our attention from the organisms. As enunciated by Raven (1977), "the preoccupation with . . . the recognition of taxonomic units may often tend to conceal the facts upon which such classifications are based. . . . The species concept, lacking a universal definition, has very little predictive value, but provides a kind of false assurance that they have a number of key biological properties." He argues, for this reason, that species concepts may mask information and distort our perception of diversity.

Stebbins (1969a) reminds us that there are not groups of individuals in nature that can be grouped in only one way as objective, uncontestable species. Yet, at the same time, species are not purely subjective assemblages, carved out of an amorphous mass of diversity. So what are we to do? We are free to create and amend species interpretations until we have a mentally satisfying organization. If these interpretations are satisfying, they will have value; and that is about as much as we can hope for.

The Ecogenetic Species Concept

The limitations of any species concept notwithstanding, I am obliged to employ one, since this book deals with the origin and fate of species. I propose that each species has a unique way of living in and relating to the environment and has a unique genetic system—that is, that which governs the intercrossability and interfertility of individuals and populations. I will refer to this as the Ecogenetic Species Concept.

Ecogenetic Species occupy different niches by being divergent in some of the following: habitat tolerance, resource allocation, phenology, pollen delivery system (e.g., type of pollinator; self-pollination vs. cross-pollination), and reproductive system (e.g., monoecious vs. dioecious). The ecological distinctiveness of a newly arisen species (neospecies) arises primarily from selection for traits that enhance the adaptedness of neospecies in environments (physical or biotic) beyond the limits of its progenitor. Selection promotes a shift in the mode of resource acquisition and utilization. Stochastic processes are unlikely to render such alterations.

I do not mean to suggest that an Ecogenetic Species has a uniform set of ecological properties. Indeed, properties of Ecogenetic Species may vary substantially among populations. Species have realms of ecological variation; and the realm of one species will be different from that of another, even if they overlap in part. Thus the ecological properties of Ecogenetic Species are not equivalent to the taxonomic properties of species, which are chosen because they are conservative and stable attributes.

In an essay on the biological meaning of species, Mayr (1969) describes some fundamental attributes of species. He writes that species "are the entities which specialize, which become adapted, which shift their adaptation. And speciation . . . is the method by which evolution advances." It is easy to see how this idea may be applied to Ecogenetic Species because the key to their origin is a shift in ecological relationships.

Each species has a distinctive genetic system that is shaped by selection on genes relating directly or indirectly to fitness, the hitchhiking of other genes, and random genic and chromosomal changes. By virtue of their distinctive genetic systems, species will differ in pollen-pistil recognition signals, developmental programs, and/or chromosomal arrangements (cryptic or major) or numbers. The ensuing differences in the ecological constitutions and genetic systems of Ecogenetic Species will reduce or preclude their ability to exchange genetic material with other species. The specific factors operative in this respect are listed in table 1.1. The relative importance of these factors may vary among species. These factors are not properties of single species; rather they reside at the hypothetical interface between species and are not independent of that interface.

Why are two criteria employed for species delineation? The ecological criterion affords a facile recognition of species' idiosyncrasies. However, ecological disparity alone is insufficient reason for according species status because it would not necessarily maintain species' integrity in areas of sympatry.

The criterion of unique genetic systems provides the basis for species' genetic independence. However, it alone is not sufficient for species recognition. Divergence in genetic systems may be only the result of stochastic processes, as in the fixation of novel chromosomal rearrangements. Populations that differed by a number of translocations could have very similar gene frequencies and could share ecological attributes, morphologies, and so on. Yet they would have to be accorded species status using only the genetic system as the guideline.

Based on the Ecogenetic Species Concept, the ecological relationships and genomic compatibilities may differ substantially among pairs of progenitors

---

Table 1.1. A classification of factors restricting or precluding interspecific gene exchange.

Ecological: species relate to resources in divergent ways which limit their opportunities to exchange pollen
   A. Habitat divergence: species occupy different habitats
   B. Temporal divergence: species flower at different times
   C. Floral divergence: species attract different sets of pollinators because they differ in floral attractants or morphology
   D. Reproductive mode: species employ self-fertilization and asexual reproduction to various degrees
Genomic: pollen is transferred between species but gene exchange between them is limited
   A. Cross-incompatibility: the pollen of one species does not send tubes to the ovules of another species or the cross is ineffective
   B. Reduced fitness of F_1 hybrids:
      (1) Hybrid inviability and weakness
      (2) Hybrid floral isolation
      (3) Hybrid sterility
   C. Hybrid breakdown in advanced or backcross generations: reduced fitness as in F_1 hybrids

and neospecies or among pairs of closely related species. Species may fall anywhere within the boundaries of ecological divergence and genomic incompatibility that meet the species criteria. As will be discussed in a later chapter, some species are quite divergent ecologically, but maintain the ability to exchange genes with congeners. At the other extreme, other species pairs are only moderately divergent but strongly incompatible.

Whereas the proposed species concept holds that related species would retain their integrity in the areas of sympatry, the possibility of hybridization and introgression is not precluded. Indeed, good Ecogenetic Species may hybridize.

Even long-term hybridization does not necessarily lead to the breakdown of species integrity, as demonstrated in *Populus*. Natural hybrids between species of cottonwoods and balsams are frequent throughout their large area of sympatry. Eckenwalder (1984) has shown that these hybrids have been produced frequently over the past 12 million years. Contemporary poplar hybrids have reduced fertility and viability relative to their parents. Most hybrids in the field appear to be first generation. Hybrids show little tendency to invade the distinctive habitats of their parents.

The Application of the Ecogenetic Species Concept

The Ecogenetic Species Concept has utilitarian value, which is important for an operational species concept. Most of the assemblages (taxa) that plant taxonomists consider species have distinctive morphological and ecological features, and are at least partially reproductively isolated from their closest relatives. Thus most Taxonomic Species are likely to be equivalent to Ecogenetic Species in their circumscription.

In many genera we can find assemblages of populations that defy classification, species concept notwithstanding. The most difficult situations often involve assemblages in which (1) hybridization is extensive, (2) populations differ in chromosome number or chromosome arrangement, and (3) reproduction is by asexual means. The application of the Ecogenetic Species Concept is problematic in the latter two cases.

Consider first a case of extensive hybridization that caused problems in species diagnosis. Sympatric species of white oaks typically have divergent edaphic requirements, yet they hybridize extensively and hybrids are partially fertile (Grant, 1981). Grant refers to groups of hybridizing white oaks as syngameons, which he defines as "the most inclusive unit of interbreeding in a hybridizing species group." Because of their promiscuity, Grant demotes taxa typically accepted as species to semispecies. Analyses of nuclear markers show that local introgression occurs at some sites, but marker exchange usually is not extensive (Guttman and Weigt, 1989; Ducousso, Michaud, and Lumaret, 1993; Howard et al., 1997; Bacilieri et al., 1993, but see Whittemore and Schaal, 1991). In general there is an abrupt genetic and phenotypic discontinuity between hybridizing species beyond the sites of hybridization. Because these species retain their integrity in the face of hybridization, they conform to the criteria of Ecogenetic Species.

Turning to chromosomal discontinuities, autotetraploids and their diploid progenitors would be conspecific under the proposed species concept, unless the former had distinctive ecological attributes. Thus differences in genome number would not be sufficient grounds for species recognition. However, autotetraploids usually have distinctive ecological properties, which are evident in their distinctive distributions and in their distinctive physiological tolerances. Accordingly, most autotetraploids would be recognized as different species. The substantial literature on this subject has been reviewed by Lewis (1980) and Levin (1983a).

Consider the case of *Tolmiea menziesii*. Diploid populations range from the southern limit of the species in California northward to central Oregon, where they are replaced by tetraploid populations that extend northward through British Columbia into southern Alaska. The cytotypes, which are morphologically very similar, have very similar allozyme frequencies (Soltis and Soltis, 1989). The tetraploids are more heterozygous than their diploid counterparts (0.23 vs. 0.07) and have alleles at some loci not present in the diploids. Diploid and tetraploid populations also possess 5S and 18S–25S ribosomal RNA genes of identical repeat length and restriction profile (Soltis and Doyle, 1987). Diploid and tetraploid populations also share identical floral anthocyanins, foliar flavonoids, and karyotypes and chromosomal banding patterns (Soltis and Soltis, 1989). Given these considerations and the fact that *Tolmiea* is a monotypic genus, the tetraploids almost certainly are autotetraploids. Ostensibly they are of relatively recent origin. Following the criteria of the Ecogenetic Species Concept—that is, ecological disparity and at least partial postpollination isolation—they would be recognized as distinct species.

As autotetraploids age, they would be expected to become more distinct, and thus more easily recognized as species. Of course, then, their autopolyploid nature becomes less apparent. This seems to be the case with the tetraploid, sandstone endemic *Solidago albopilosa* (Esselman and Crawford, 1997). This species shares many allozyme alleles with the diploid *S. flexicaulis*, but has some unique ones as well. The species are morphologically similar but readily distinguished. They are ecologically disparate (*S. flexicaulis* grows in woodlands) and have postzygotic barriers related in part to ploidal differences. Accordingly, *S. albopilosa* is readily assigned Ecogenetic Species status.

The proposed species concept recognizes as heterospecific two populations that differed in major chromosomal rearrangements and in ecological tolerances and strategies. Consider *Clarkia rhomboidea*, in which six translocations occur among 26 populations (Mosquin, 1964). Populations are homozygous for these translocations, and interpopulation hybrids have much reduced fertility. Some chromosomally divergent populations have dissimilar ecological requirements and morphologies; others do not. The application of the Ecogenetic Species Concept would be a taxonomic nightmare, but so would the application of the other concepts as well.

The application of the Ecogenetic Species Concept also may be problematic in its application to ecogeographically and morphologically distinct population systems that replace one another along environmental gradients or that simply have different habitat requirements. Consider *Potentilla glandulosa* in

California. Clausen and Hiesey (1958) describe four subspecies (*typica, reflexa, hanseni,* and *nevadensis*) that replace each other as one moves from coastal areas through climatic zones of the Sierra Nevada Range. The subspecies hybridize; yet they remain distinct. Hybrids are fertile and often are as vigorous or more vigorous than the parental taxa in the field and in garden trials. Advanced generation hybrids are readily obtained. Given the absence of significant post-mating barriers (or discordance in genetic systems), the four subspecies would fall within one Ecogenetic Species.

The situation is not always so simple. Ecologically distinctive entities may have moderate to substantial postmating barriers as in the *Strepthanthus glandulosus* complex, where hybrid fertility is a negative function of distance in some taxa (Kruckeberg, 1957). A similar pattern occurs in the *Gilia achilleaefolia* complex (Grant, 1954) and in species of *Epilobium* (Raven and Raven, 1976). In other species there is no simple geographical pattern to the strength of barriers between ecologically distinctive entities (e.g., *Mimulus guttatus* and *M. glabrata* complexes Vickery, 1978).

The application of the Ecogenetic Species Concept makes sense in groups where taxonomic treatments reflect morphological and ecological differences, on the one hand, and postmating barriers, on the other hand. For instance, the *Strepthanthus glandulosus* complex contains five taxa that had been considered by some to be one species. Kruckeberg (1957) treats two taxa with strong postpollination barriers as separate species, while the remaining entities fall within *Strepthanthus glandulosus*. In the absence of these barriers, Kruckeberg would recognize one species as would the Ecogenetic Species Concept.

Serpentine-tolerant and intolerant populations occur within *Strepthanthus glandulosus* (Kruckeberg, 1957). Some crosses between tolerant and intolerant populations yield hybrids with low fertility, but so do some crosses between tolerant populations and some crosses between intolerant populations. Given that the two groups are not predictably different in genetic system harmony, there would be no grounds to recognize them as distinctive Ecogenetic Species. Kruckeberg's taxonomic treatment is consistent with this diagnosis.

Overview

There is a seemingly incessant discussion about species concepts, and an occasional formulation of a new concept. This speaks to the different perspectives of the practitioners and to the dissatisfaction with existing concepts. The choice of a species concept has to do, in part, with the perspective that gives one satisfaction. Many practitioners hold that reproductive isolation is the key to species delimitation, and as such their view of speciation involves the origin of these barriers. Others hold that ecological disparity is the key to species delimitation. Thus, in the former speciation may have little to do with adaptation, whereas in the latter speciation involves adaptive change.

Each species concept has a unique set of organizing principles that lead to a unique way of partitioning diversity. When groups of populations are divergent among themselves on ecological, genetical, and phenotypic grounds, then the application of different species concepts are likely to yield the same species interpretations.

The application of each species concept has its limitations. Some work better in one group of species than in another, even though the groups are in the same genus. Clusters of distinctive populations do not sort themselves out into well-mannered, discrete packages no matter how we circumscribe them. Thus it is important to understand the biological properties and relationships of population systems rather than getting hung up on putting them into conceptual pigeonholes.

2

Speciation

The Ecological Transition

The role of ecological processes in speciation has been the subject of debate for several decades (Schluter, 1998). Two principal ideas emerge. The most prominent is that divergent natural selection on populations in different environments drives the development of pre- and postzygote barriers to gene exchange (Mayr, 1942, 1963; Dobzhansky, 1951). The second idea is that ecological processes influence speciation principally through their effect on the viability of populations (Mayr, 1963; Allmon, 1992). How these processes work will be the subject of much of this book.

Speciation Trajectories

Studies of recently evolved species (neospecies) and their progenitors reveal that speciation may follow any one of a multitude of trajectories (figure 2.1). At one extreme, new species would be very different in the acquisition and utilization of resources, while being only partially isolated by postpollination barriers. Ecological change in this scenario is the primary element in speciation. At the other extreme, new species would be moderately distinct in their ecological profiles, but have strong postpollination barriers. Here barrier building is the primary element in speciation. Multiple trajectories are possible because the levels of ecological divergence and genomic incompatibility are not necessarily correlated.

Speciation with an Ecological Emphasis

Tetramolopium and the silverswords in Hawaii followed the general trajectory involving considerable ecological change with modest postmating isolation.

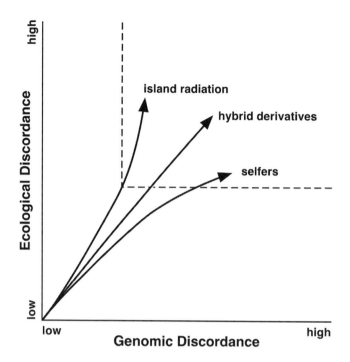

Figure 2.1. Possible trajectories of speciation. Species status is reached when arrows penetrate the dashed line.

The silverswords (*Argyroxiphium, Dubautia, Wilkesia*) occupy nearly the entire range of environmental conditions in the islands (Carr and Kyhos, 1986; Baldwin and Robichaux, 1995). They occur from 75 to 3,750 m in elevation in sites that differ in annual precipitation from less than 400 mm to over 12,000 mm. Species occur in dry shrublands, dry forests, subalpine shrublands and forests, alpine deserts, and mesic and wet forests. *Tetramolopium* species are distributed from near the coast to 3,300 m, and reside in habitats ranging from coastal shrubland to grassy flats, cliff faces, dry forests, subalpine woodlands, and alpine deserts (Carr and Kyhos, 1986; Baldwin and Robichaux, 1995).

Congeneric species in the two groups are interfertile, and so are some silverswords belonging to different genera. In the silverswords, pollen fertility of interspecific hybrids within a genus ranges from ca. 30 to close to 100%. Intergeneric hybrids have pollen fertilities from ca. 10 to ca. 30%. First-generation hybrids between *Argyroxiphium sandwicense* and *Dubautia menziesii* are weakly fertile (9% pollen stainability; Carr, 1998). However, first-generation backcrosses involving the latter are highly fertile (79% pollen stainability) as are second-generation backcrosses (99% pollen stainability). In *Tetramolopium* most of the hybrid combinations are highly fertile. Pollen fertility generally is greater than 80%, and F_1, F_2, and F_3 hybrids are viable and fertile in all possible combinations among species (Lowrey, 1995).

The genus *Espeletia* has radiated into numerous species in high montane forests of the northern Andes. Species vary in habit from herbs to trees over 20 m tall. Achene-set from interspecific crosses exceeds 50% in most crossing combinations, and germination is high, even though species are morphologically divergent (Berry and Calvo, 1994). The fertility of these hybrids is unknown, but the presence of natural hybrid swarms indicates that hybrids are at least partially fertile.

Species of *Epilobium* in New Zealand are another assemblage in which speciation is accompanied by manifest ecological divergence. Yet most species are readily crossed, and hybrids of most species combinations are quite fertile (Raven, 1972).

There are many examples of edaphic specialization associated with speciation. The narrow serpentine endemic, *Layia discoidea*, is derived from the soil generalist, *L. glandulosa* (Gottlieb, Warwick, and Ford, 1985). This change has been accompanied by striking changes in floral and vegetative morphology. In spite of the edaphic and morphological divergence, the species cross readily in the greenhouse and hybrids are fully fertile. Their genetic identity is 0.90, which is only slightly lower than that between conspecific populations.

Edaphic change without the emergence of strong postpollination barriers occurred during the evolution of *Lasthenia maritima* from *L. minor* (Crawford, Ornduff, and Vasey, 1985). The latter grows in a variety of habitats in mainland California and is an outcrosser. Its derivative, which is a selfer, occurs on seabird rocks whose substrate is heavily manured, nitrogenous, and acidic. Interspecific hybrids have high fertility. The genetic identity of the species is 0.89, which is nearly as high as the identity of conspecific populations.

The type of ecological shifts that promoted speciation may be inferred from comparisons of terminal pairs of species in cladograms. This is best done when sister species are quite similar in their molecular sequences, which signals recent divergence. For example, in *Lapeirousia* subg. *Lapeirousia* (Iridaceae), speciation usually is combined with specialization for different soil types and/or pollinators (Goldblatt and Manning, 1996). Edaphic change has been from coarse and sandy soils to clay soils. Pollinator use has shifted from long-tongued flies and moths to bees and generalist pollinators. Where both shifts occurred, the edaphic shift is thought to have preceded the biotic.

Speciation with Emphasis on Genomic Change

Speciation may involve moderate ecological divergence accompanied by the development of strong postpollination barriers, and thus follows a trajectory that is the converse of that previously discussed. Populations subject to stochastic processes associated with colonization or population contractions may fix novel chromosome arrangements, or cytoplasmic haplotypes and nuclear genes that confer cross-incompatibility or hybrid inviability. These populations also may fix unusual phenotypic traits. Should isolated populations be subject to severe environmental stresses, they might respond to selection for traits enhancing tolerance to the unusual environment. The origin of many *Clarkia* neospecies ostensibly involved this progression (Lewis, 1962, 1973; Vasek,

1964, 1968, 1977). Conspecific populations that are divergent in chromosomal arrangements are well documented in the genus (e.g., *C. rhomboidea*, Mosquin, 1964).

Another example of speciation involving considerable chromosomal change and moderate ecological change is in *Coreopsis*. From allozyme variation, Crawford and Smith (1982) showed that *C. nuecensis* is a likely derivative of *C. nuecensoides*. The genetic identity of the species is 0.97. The former has fewer chromosomes than its progenitor, and interspecific hybrids are almost completely sterile, though vigorous. The species occupy somewhat different soil types. Both species are annual and have similar morphologies.

Where ecological change is the major factor in speciation, ecological change ostensibly precedes postzygotic barrier building. The converse is likely to be the case where the emergence of postzygotic barriers is the predominant factor in speciation. Chromosomal rearrangements involve stochastic processes, whereas ecological shifts involve directional selection. A small population ($\leqslant 100$) does not offer a favorable substrate for selection; and a large ecologically attuned population does not offer a favorable substrate for the fixation of novel chromosomal variants.

Speciation Without Emphasis

Having described the two extreme speciation trajectories, consider one of the numerous possible intermediate speciation trajectories in which ecological and genetic system transitions are both substantial and more or less concurrent. One example is in *Gaura*. The origin of *G. demareei* from *G. longiflora* involved a shift in reproductive ecology (Gottlieb and Pilz, 1976). Flowers of *G. demareei* are larger, open at sunrise, and are pollinated by bees. Flowers of *G. longiflora* open at sunset and are pollinated by moths. The species differ by at least two reciprocal translocations and are partially sterile. Speciation must have been very recent as their genetic identity is 0.99, which is the same as among conspecific populations.

Speciation via hybridization and chromosome doubling follows an intermediate trajectory. The allopolyploids are apt to have reduced cross-compatibility with their progenitors, and hybrids tend to be sterile (Levin, 1978a). Allopolyploids typically have distinctive ecological tolerances relative to their progenitors as is often indicated by their geographical distributions.

The basis for an ecological shift in polyploids comes in part from a change in ploidal level per se (Levin, 1983a). Autopolyploids may be more tolerant of nutrient-poor soils than their diploid counterparts, in part because of greater nutrient uptake efficiency. Autopolyploids may be more drought-resistant. Chromosome doubling often alters resistance to cold, but the direction of response varies among species. Chromosome doubling also may alter a plant's resistance to pests and pathogens through changes in the character and concentration of secondary products.

In addition to the effect of ploidal change per se, allopolyploids are likely to have ecological tolerances differing from their parents because they contain a novel combination of genes and increased heterozygosity. They may be able to mobilize genetic variation more effectively than their diploid progenitors through intergenomic recombination and through higher mutation rates (Soltis and Soltis, 1995). Gene silencing and genetic diversification, which result in the functional or regulatory divergence of duplicate genes, also may allow allopolyploids to evolve novel ecological tolerances.

Recombinational speciation—that is, the stabilization of the diploid products of hybridization—also follows an intermediate trajectory with regard to the development of new ecological tolerances and new postpollination barriers. This has been very well demonstrated in *Helianthus* (Rieseberg, 1991a; Rieseberg, Carter, and Zona, 1990; Rieseberg, Van Fossen, and Desrochers, 1995; Rieseberg et al., 1996). The parental species, *H. annuus* and *H. petiolaris*, are restricted to heavy clay soils, and to dry sandy soils, respectively (Rieseberg et al., 1995). Nevertheless, the species are sympatric throughout the western United States and hybrid swarms are common. Two diploid hybrid species, *H. anomalus* and *H. deserticola*, are endemic to habitats more xeric than tolerated by *H. petiolaris*. Another stabilized derivative of the same parentage, *H. paradoxus*, is narrowly distributed in brackish or saline marshes. The hybrid derivatives have strong postpollination barriers among themselves and with the parental species.

The biological relationship between progenitor and derivative depends on which trajectory was taken. If the evolution of new tolerances (adaptations) occurred first, then obstructions to interspecific gene exchange due to cross-incompatibility, hybrid inviability, and hybrid sterility are likely to be minor, especially in the early history of a species. The ability of the neospecies to retain its genetic integrity when sympatric with the progenitor would be due largely to prepollination barriers. The breakdown of these barriers would lead to wholesale hybridization. If chromosomal rearrangement or other postpollination impediments to gene exchange developed first, complex hybrid swarms involving the progenitor are unlikely to develop.

The Evolutionary Lability of Characters

Given that ecological change is a prerequisite for speciation, we need to know the evolutionary lability of ecologically relevant traits such as pollination system, breeding systems, and habitat tolerance. Can populations readily alter these traits when new opportunities (selective pressures) are presented? The answer to this question can be obtained directly from selection experiments where response is measured on a generation by generation basis or from estimates of heritability. We must keep in mind, however, that gene flow and negative genetic correlations may substantially retard the response to selection in the real world (Antonovics, 1976).

If we take a broader perspective, the lability of ecologically significant traits may be inferred from phylogenetic studies. High levels of diversification within a genus and the multiple origins of character states in different congeneric species are indicative of evolutionary lability. Historical inferences concerning character evolution are best made when well-resolved phylogenies, particularly at the species level, are available.

Pollination Systems

A shift in pollinators is one of the prime factors responsible for diversification in floral size and form, pigmentation, and rewards within many plant genera (Grant and Grant, 1965; Stebbins, 1974). Indeed, if plant-pollinator relationships are specialized, shifts in pollinators favor an alteration in the floral syndrome of populations and thus promote speciation (van der Pijl, 1961; Baker and Hurd, 1968). Floral syndromes are suites of traits that often reflect adaptations to one group of pollinators or another (e.g., birds vs. bees vs. lepidopterans).

Change in a pollinator niche is possible if populations contain novel floral variants that new pollinators better exploit or find more attractive. Most shifts are likely to be the direct result of selection, but some may be correlated responses to selection on traits other than flowers, and others may be the result of genetic drift (Wilson and Thomson, 1996). Most shifts in the category of pollinator would be accompanied by phenotypic change and would afford some measure of reproductive isolation.

Evolution in response to selection by pollinators requires (1) phenotypic variation in floral or inflorescence architecture or floral rewards and signals, (2) a relationship between a fitness component related to pollination and a trait value, and (3) a heritable component to phenotypic variation (Endler, 1986). Natural selection for floral traits has been inferred in many species from the differential success of alternate phenotypes in seed production or pollen export. For example, in *Ipomopsis aggregata*, hummingbirds favor wider corollas, which in turn promotes greater pollen export to other plants because hummingbirds can remove a greater fraction of the pollen presented on the anthers of wide corollas (Campbell et al., 1991; Campbell, Waser, and Price, 1996). With an estimate of heritability for this trait (0.3), Campbell was able to make predictions about the response to selection in nature (0.03–0.04 mm per generation). The mean corolla width is ca. 3.6 mm. Longer corolla length also promotes pollen export in *I. aggregata*. The predicted response to selection is 0.04–0.13 mm per generation. The mean corolla length is ca. 25.0 mm.

In order to be of any consequence, reproductive differentials between phenotypes must translate into a shift in trait expression in the next generation. Galen (1995, 1996) demonstrated that bumblebees exert selection for broadly flared corollas in *Polemonium viscosum*. Flare (lobe) length varies from 8 mm to 18 mm. Seed-set is significantly correlated (positively) with flare length, as is pollen export. Corolla flare averages 14.4 mm in the progeny of bee-pollinated parents vs. 13.1 mm in hand-pollinated control progeny (figure 2.2). The 9% increase in flare after one generation of selection is statistically significant.

The literature contains numerous estimates of heritability conducted in the greenhouse. Unfortunately, such studies overestimate heritability in nature, because while the genetic component of phenotypic variance is about the same in the field and in the greenhouse, the environmental component of phenotypic

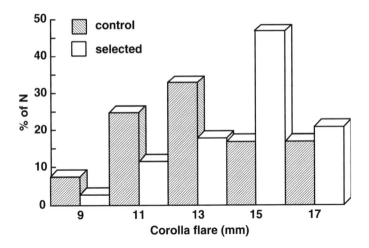

Figure 2.2. Frequency distribution of corolla flare of bumblebee selected plants and progeny of hand-pollinated controls in *Polemonium* (from Galen, 1996. Redrawn with permission of *Evolution*).

variance is much lower in the greenhouse. Only a few studies have been conducted in the field. These studies show that some traits have the potential to respond to selection, at least in the study populations. For example, narrow-sense heritability for corolla length is 0.14 in *Phlox drummondii* (Schwaegerle and Levin, 1991), and between 0.24 and 0.74 in *Ipomopsis aggregata* (Campbell, 1996). Corolla width, stigma position, and anther position also have moderate heritabilities in *Ipomopsis*. In *Polemonium viscosum* heritability for corolla flare is not significantly different from 1.00 (Galen, 1996).

In addition to being dependent on trait heritability, floral evolution depends on the genetic correlations between traits. Negative correlations between traits related to fitness may retard the response to selection. Conversely, positive correlations may facilitate multiple character evolution. Estimates of genetic correlations are environment-dependent, as are estimates of heritability. Campbell's (1996) field study of *Ipomopsis aggregata* showed that corolla length is positively correlated with corolla width and anther and stigma positions.

Shifting to a phylogenetic perspective, we find that pollination systems may be subject to change over a relatively short period of time with repeated parallelisms and reversals. Perhaps the classic study on floral radiation in plants was Grant and Grant's (1965) work on the Polemoniaceae. They showed that bee, lepidopteran, fly, and hummingbird pollination syndromes have evolved independently in different congeneric species and different genera (Grant and Grant, 1965).

Examples of repeated evolution emerge when pollination systems in *Dalechampia* (Euphorbiaceae) are mapped on a modern phylogeny. Species with flowers that secrete monoterpene fragrances and are pollinated by fragrance-collecting male euglossine bees originated independently three or more times from species whose flowers secrete triterpene resins and are pollinated by female bees that collect the resin (Armbruster, 1993). The sites of fragrance secretion and the likely biochemical pathways involved were quite different among lineages.

In *Lapeirousia* subgenus *Lapeirousia* (Iridaceae), there are two long-tongued fly pollination syndromes that differ principally in tepal pigmentation (Goldblatt, Manning, and Bernhardt, 1995). The *L. silenoides* type has evolved independently in four species and the *L. fabricii* type in three species. Bee or generalist pollination has evolved from species catering to long-tongued pollinators four or five times in conjunction with a shortening of the flower tube.

In the southern Africa orchid genus *Disa*, butterfly-pollinated flowers have arisen twice, night-scented flowers pollinated by moths, three times, long-spurred flowers pollinated by long-tongued-flies, four times, and showy, deceptive flowers pollinated by carpenter bees, twice (Johnson, Linder, and Steiner, 1998). Red flowers have evolved twice, and nectar has evolved three times. In the tropical legume genus *Erythrina*, shifts from passerine to hummingbird pollination have occurred a minimum of four times (Bruneau, 1997).

The transition from animal pollination to wind pollination has been made only in a few genera. Yet this transition has occurred independently at least four times in the Hawaiian genus *Schiedea* (Caryophyllaceae; Weller et al., 1998). Of particular interest is that the transition is associated with the evolution of dioecy, which involves changes in pollen size, pollen production per flower, pollen:ovule ratio, and nectar concentration (figure 2.3). This shift accompanied the colonization of dry habitats by plants from mesic habitats.

Stebbins (1974) asserted that shifts between pollination systems involve an intermediate phase in which both new and old systems operate. Armbruster (1993) tested this assertion in *Dalechampia* and identified quantitative shifts, qualitative shifts with an intermediate phase, and qualitative shifts without an intermediate phase. Quantitative shifts occur when a change in the mean value of character such as nectary gland size, gland-stigma distance, and anther-gland distance is accompanied by a change in the relative importance of different pollinator species. Qualitative shifts are manifested in the kinds of rewards that attract pollinators (resin or fragrance) and differences in resin biochemistry. Quantitative change may be gradual, whereas qualitative change may be abrupt. Both changes produce ethological barriers to pollen exchange because pollinators may be highly specific in their foraging behavior.

A shift from one pollination system to another or from one reproductive system to another requires several generations of selection in one direction by the pollinator fauna. Variation in the intensity would not negate the process, whereas variation in the direction of selection may.

It has long been assumed that the pollinator fauna is a consistent agent of selection. This idea recently has been challenged by Waser et al. (1996), Herrera (1996), and Ollerton (1996) on two counts. First, recent studies show that plants may experience wide variation in abundance and composition of pollinators in time and space. The stochastic nature of spatio-temporal variation in pollinator composition within and among years and differences in effectiveness

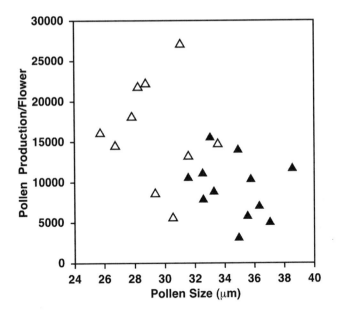

Figure 2.3. Pollen production per flower versus pollen size for *Schiedea* species. Closed triangles designate biotically pollinated species; open triangles designate wind-pollinated species (from Weller et al., 1998. Redrawn with permission of the Botanical Society of America).

of pollinator species results in an inconstant selection regime for plants. Second, surveys of plants and their pollinators show that plant-pollinator relationships are usually not as specific as might be inferred from the floral traits (syndrome) of a species. Indeed there is substantial evidence that syndromes have little predictive value in explaining interspecific differences in pollinator composition. Moreover, the floral attributes of many species may have little to do with contemporary pollinators. Where this is the case, the importance of pollinators in the isolation of species will be less than we might imagine.

Inconsistent selection on floral traits has been documented in two populations of the hummingbird-pollinated *Lobelia cardinalis* (Johnston, 1991). The population in which seed-set is less limited by pollen experiences positive directional selection on flower number and median-flower date, whereas the other population experiences positive directional selection on nectar-stigma distance and inflorescence height, as well as flower number. The heterogeneous selection pressure in *Lobelia* may be related to differences in the timing of peak pollinator service.

Inconsistent selection on floral traits in time has been found in *Calathea ovandensis* (Marantaceae; Schemske and Horvitz, 1989). Shorter corolla tubes were favored in one year of three. The heterogeneous selection pressure in

Calathea apparently is due to differences in the abundance of the most efficient pollinator.

Even if consistent or nearly consistent selection on floral traits is not the norm, it does occur. For example, Campbell et al. (1991) demonstrated selection for wider corolla width in six of seven populations of *Ipomopsis aggregata*. Perhaps consistent selection mostly occurs in isolated populations where the normal pollinator fauna is replaced by one pollinator group that is abundant over many years (Ollerton, 1996). Long-term studies on plant-pollinator relationships within single populations are needed to test the validity of this hypothesis.

Breeding Systems

One line of evidence that breeding systems are labile lies in the many self-compatible domesticates that have been derived from self-incompatible wild progenitors (Rick, 1988). Some domesticates are vegetable crops (e.g., tomatoes, endives, cauliflower); others are ornamentals (snapdragon, phlox, petunia). An increase in the penchant for selfing usually has been accompanied by an increase in the proximity of anthers and stigmas and the synchrony of anthesis and stigma receptivity.

Another line of evidence that breeding systems are labile lies in the evolution of predominantly selfing populations from outcrossing populations in a multitude of wild plant species. For example, *Gilia achilleaefolia* (Polemoniaceae) contains xenogamous and autogamous races that differ substantially in their outcrossing rates (Grant, 1954; Schoen, 1977, 1982). Self-pollinating populations also occur in the typically cross-pollinating *Arenaria uniflora* (Caryophyllaceae; Wyatt, 1988) and *Clarkia xantiana* (Onagraceae; Moore and Lewis, 1965).

The results of artificial selection on wild populations also speak to the lability of breeding systems. Consider *Phlox*, where selection was practiced for increased and decreased autogamy (Bixby and Levin, 1996). Two cycles of selection in the predominantly outcrossing *P. drummondii* increased autogamous seed production from 4 to 56% of the ovules in one population, and from 22 to 41% in another (figure 2.4). Two cycles of selection for decreased autogamy in the related and predominantly selfing *P. cuspidata* also had a significant response. In one population autogamy declined from 84 to 69%, and in another it declined from 82 to 53%.

It is noteworthy that even cultivars respond to selection on autogamy, although they have less overall genetic variation than populations of their wild progenitors. Two cycles of selection for autogamy in the *Phlox drummondii* cultivar 'Salmon Beauty' increased autogamous fruit-set from 40 to 95%; two cycles of selection against autogamy decreased autogamous fruit-set from 40 to 12% (Bixby and Levin, 1996). This cultivar has half the proportion of polymorphic loci (0.11 vs. 0.21), one-fourth the level of heterozygosity (0.01 vs. 0.04), and one-third the genetic diversity (1.24 vs. 3.49) of wild *P. drummondii* populations based on allozyme analyses (Levin, 1976b).

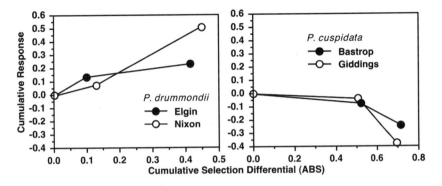

Figure 2.4. Cumulative response to selection on proportion autogamous fruit-set as a function of the absolute value (ABS) of the cumulative selection differential in *Phlox drummondii* (from Bixby and Levin, 1996. Redrawn with permission of *Evolution*).

The breeding system also was altered in opposing ways in turnips (Richards and Thurling, 1973). One cycle of selection for self-fertility raised self-seed from ca. 2 per fruit to ca. 5. One cycle of selection for self-infertility brought self-seeds/fruit down to 0.74.

The evolutionary lability of breeding systems is evident when system traits are mapped onto estimated phylogenies of a genus. In the Polemoniaceae, autogamy has evolved independently from xenogamy at least 14 times, nine of which occurred in the *Gilia-Allophyllum* (Polemoniaceae) clade, which has homomorphic incompatibility (Barrett, Harder, and Worley, 1996). The phylogenetic tree for *Gilia* and the independent origins of autogamy is depicted in figure 2.5.

Xenogamy also breaks down in genera with heteromorphic incompatibility. Distyly is considered to be the primitive condition in *Amsinckia* (Boraginaceae; Schoen et al., 1997). There appears to have been four separate transitions from distyly to homostyly, and thereby from xenogamy to autogamy. In *Eichhornia* there were at least three transitions from tristyly to homostyly (Kohn et al., 1996).

Breeding system evolution can be reversible. The best example is in the neotropical vine *Dalechampia*, where autogamous species have evolved from more allogamous ones up to 11 times; and relatively allogamous species have evolved from more allogamous species up to 13 times (Armbruster, 1993).

In general the transition from predominantly outcrossing to predominantly selfing species involves an increase in pollen-pistil compatibility and alterations in floral attributes that facilitate self-pollination, on the one hand, and reduce the cost of reproduction, on the other. This transition may be less plagued by inconsistent selection than the transition from one pollinator group to another. Selfing is favored when there is a sustained paucity of pollinators servicing the species such that seed-set is substantially below the norm (Jain, 1976). However, the transition to selfing typically is opposed by inbreeding depression (Charlesworth and Charlesworth, 1987).

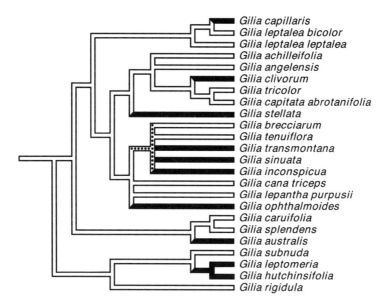

Figure 2.5. The origins of autogamy in *Gilia*. The white branches denote xenogamy, black branches autogamy, stippled branches undetermined (from Barrett et al., 1996, Fig. 3. Redrawn with permission of the Royal Society of London).

Flowering Time

Flowering time is another reproductive attribute that is responsive to selection. Intraspecific variation in flowering phenology has been demonstrated in numerous species when plants from different localities are grown together in the greenhouse or in the field (Heslop-Harrison, 1964; Langlet, 1971). In general plants from higher altitudes or higher latitudes flower earlier than those from lower altitudes and latitudes. Plants from closely adjacent populations also may differ in flowering time as a result of natural selection. For example, in the graminaceous species *Agrostis tenuis* and *Anthoxanthum odoratum*, mine populations flower about a week earlier than pasture populations (McNeilly and Antonovics, 1968). This may be the result of selection for more rapid maturation on the drier mine sites or selection on the mine plants to avoid gene flow from pasture plants.

Artificial selection on flowering time rarely has been conducted on wild populations. Carey (1983) conducted five generations of selection for earlier and later flowering in the outcrossing *Plectritis congesta* (Valerianaceae) and the predominantly self-fertilizing *P. brachystemon*. In *P. congesta* the early line flowered 12 days earlier than the control, and the late line 17 days later than the control (figure 2.6). The response in the selfer was less, but still substantial (7 days earlier than the control in the early line and 14 days later in the late line). Jain (1977) practiced two generations of selection for earlier and later flowering in *Limnanthes alba*. The early line flowered an average of two days earlier than

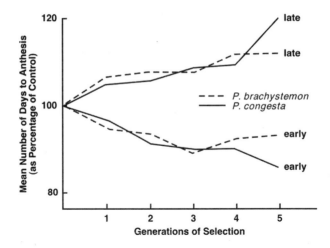

Figure 2.6. Mean number of days to flowering in populations selected for flowering time in *Plectritis* (from Carey, 1983. Redrawn with permission of *Evolution*).

the control, and the late line flowered an average of two days later than the control.

Divergence in flowering time was incidentally achieved by Paterniani (1969) when he selected against hybridization between white flint and yellow sweet maize. After five generations hybridization declined from 42 to 4%. Isolation occurred through the reinforcement of a weak phenological barrier to the point where the white flint flowered five days earlier than the original material and the yellow sweet two days later (figure 2.7).

Nonreproductive Attributes

The evolution of the annual habit from the perennial habit is a recurring theme in many plant genera. One of the best illustrations of the evolutionary lability of habit is in the Polemoniaceae. The annual habit has evolved at least seven times within the temperate clade (a tropical clade is basal); and reversion to the perennial habit apparently has occurred three times (Barrett et al., 1996). Recurrent evolution to the annual habit was primarily among desert species.

Life form is another attribute that has evolved in a multitude of genera. The evolutionary lability of life form is nicely demonstrated in the composite genus *Montanoa*. Some species are shrubs, others are trees, and others are vines. Mapping the growth form of 30 taxa on a phylogenetic tree, Funk (1982) discovered that the shrub form was pleisiomorphic (a trait exhibited by the common ancestor), and that the tree form had evolved independently four times, and the vine form three times. Species with the tree form are members of different clades, and in each case their shrubby sister species live at adjacent lower elevations. The tree form represents convergent evolution resulting from

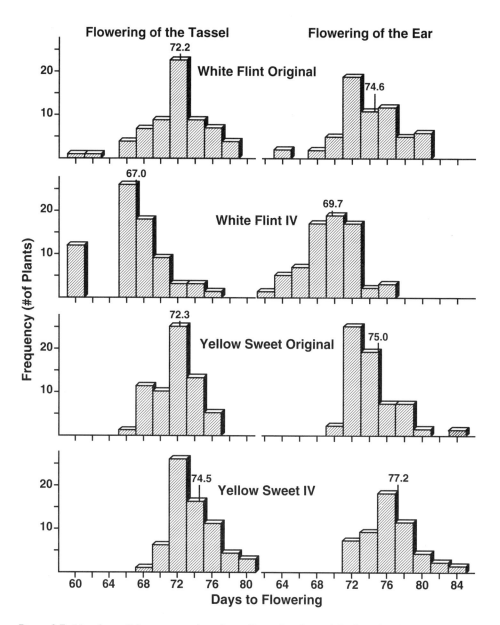

Figure 2.7. Number of days to tassel and ear flowering for original and cycle IV populations in maize (from Paterniani, 1969. Redrawn with permission of *Evolution*).

selection for survival in cloud forests. They share similar anatomical and morphological attributes. Moreover, these species are high-level polyploids.

The evolutionary lability of ecological tolerances also may be inferred from phylogenetic studies. Mayer and Soltis (1994) demonstrated the recent and independent evolution of serpentine tolerant taxa from nontolerant taxa in *Strepthanthus* (Cruciferae). Moreover, serpentine tolerance appears to have been lost repeatedly among the descendants of tolerant progenitors. Serpentine soils typically have Mg:Ca ratios greater than 1, whereas normal soils typically have more Ca than Mg (Malpas, 1991). Serpentine soils also tend to be nutrient poor and free draining, and are associated with low productivity. Thus the shift from normal soils to serpentine soils is no small achievement.

The edaphic lability of species also is evident in the polyphyletic origins of heavy metal tolerant populations. The recent, rapid, and repeated evolution of metal tolerance has been described in *Agrostis capillaris* (Al-hiyaly, McNeilly, and Bradshaw, 1988), *Agrostis stolonifera* (Wu, Bradshaw, and Thurman, 1975), *Agrostis tenuis* (Nicholls and McNeilly, 1982), and *Silene vulgaris* (Schat, Vooijs, and Kuiper, 1996).

Multiple invasions of xeric habitats by plants from mesic ancestors are known in numerous genera. In the Hawaiian *Dubautia* (Asteraceae), there have been at least four independent shifts from wet to dry habitats (Baldwin and Robichaux, 1995). Dry habitats receive less than 1,200 to 1,400 mm precipitation per year, with a conspicuous dry season. Wet habitats receive a greater amount and lack a dry season. Recurrent incursions of xeric habitats are well known in *Clarkia*, where new species have developed in ecologically marginal habitats on the periphery of species ranges (Lewis and Lewis, 1955; Sytsma, Smith, and Gottlieb, 1990). Because the phylogenetic relationships of *Clarkia* species are well understood, the adaptations of derived species can be compared with their predecessors. These adaptations include earlier maturation, lower transpiration rates, and smaller seeds and leaves (Small, 1972).

The Demography of Niche Shifting

The selective gene substitutions associated with speciation are most likely to occur in a habitat that is new for the species (Wright, 1977, 1982; Carson and Templeton, 1984; Grant, 1981; Paterson, 1985; Vrba, 1995). Entrance into and establishment in the new habitat may be achieved by temporal alterations of an existing habitat or by dispersal into a different habitat. The new habitat may differ from the standard in physical or biotic attributes. Entrance into the new habitat is facilitated when a population is pre-adapted for such or when it rapidly evolves key innovations (Futuyma, 1986; Allmon, 1992). Some members of a population have traits that are selectively disadvantageous in the traditional habitat, but advantageous in the new habitat. For example, heavy metal tolerance is advantageous on a mine site with metalliferous soils yet disadvantageous in a normal pasture site because tolerant plants grow slower than nontolerant plants (Bradshaw, 1984).

A new habitat may be regarded as ecologically marginal or stressful given that the species had not been successful there previously. Plants initially may be poorly adapted to the substrate or lack the ability to successfully interact with competitors, pollinators, herbivores, or pathogens. Thus plant growth and reproduction may be only marginally sufficient for populations to persist. The more divergent the habitat relative to the species norm the poorer will be the performance of the colonists.

Initially populations established in a new habitat are likely to be small. With low pressure from competitors and pests, populations can persist until they mobilize new adaptive phenotypes that will allow greater exploitation of the ecological opportunity and allow population growth. Otherwise these populations probably would go extinct in short order. A species makes numerous unsuccessful incursions via seed dispersal into new habitats, and only very rarely is successful (Bradshaw, 1991).

Extensive population growth may facilitate the establishment of a "platform" for a broad array of phenotypic variants, some of which would fail to reproduce in an ordinary environment (Carson, 1968). Novel phenotypes would then be subject to selection. Ultimately the changes that allow the species to shift its niche in the first place may be augmented by those "released" during the population flush.

The "exploration" of new habitats by a species will be of evolutionary consequence only if populations are restricted to and thus isolated in the new habitat, largely free of gene flow from a larger population system (Paterson, 1986). Then the gene pool can best respond to the novel selective environment as dictated by the genetic repertoire of the colonists. As Bradshaw (1984) and others have shown for heavy metal tolerance, the occupation of a new habitat (or the evolution of heavy metal tolerance) is contingent upon the requisite genes in the source populations.

Given the requisite genes, it is not difficult to imagine how the continuous input of seeds into a new habitat may eventually lead to its colonization. What is difficult to imagine is how the products of hybridization could invade a new habitat as was the case in *Helianthus*. As mentioned earlier, *H. annuus* and *H. petiolaris* are the parents of three diploid hybrid species. They are restricted to heavy clay soils and dry sandy soils, respectively. Two hybrid species, *H. anomalus* and *H. deserticola*, are more xerophytic than *H. petiolaris*, and *H. paradoxus* occurs in saline and brackish marshes.

Where did the genes for novel tolerance come from? How did a niche shift occur without a large number of seeds raining onto a new habitat? Three possibilities come to mind. First, the genes may have been present in the parental species to begin with. Artificial selection for novel habitat tolerance in both parents and in synthetic hybrid populations would shed light on the ability of populations to change in the intended direction. Second, the genes for tolerance to new habitats appeared, and a niche shift occurred, after the stabilization of the hybrid derivatives. Stabilization may have occurred in sites where species were intermixed. If the second possibility were the case, eventually there might be abundant seed available for the colonization of nonparental habitats. Otherwise, hybrid swarms would have to be very close to new habitats in order for seeds to reach them. The occasional long-distance dispersal of seed to a distant site would not suffice. The third possibility is that new gene combinations in the hybrids enabled the entrance to a new habitat. If this were the case, some contemporary hybrids (wild or synthesized) might be tolerant of habitats beyond those occupied by the parental species.

Adaptive Radiation

Oceanic island archipelagoes are especially interesting because several types of environments may be present on high elevation islands, and similar environments may be present on different islands within an archipelago. This sets the stage for adaptive radiation—that is, "the evolutionary divergence of members of a single phyletic line into a series of different niches or adaptive zones" as defined by Mayr (1970). Indeed this is what we see in many oceanic island genera (Carlquist, 1974; Crawford, Witkus, and Stuessy, 1987; Crawford and Stuessy, 1998, Wagner and Funk, 1995).

Patterns of Radiation

A likely scenario for adaptive radiation is colonization of one island in a habitat equivalent to that occupied by the source (mainland) population. Later the species spreads to other islands where similar habitats are present. Finally there is the successful invasion of "new habitats" on inhabited islands. This sequence is well documented in *Argyranthemum* (Francisco-Ortega, Jansen, and Santos-Guerra, 1996a) of the Canary Islands, and *Tetramolopium* (Lowrey, 1995), *Schiedea* (Wagner, Weller, and Sakai, 1995) and in the silverswords (*Argyroxiphium*) (Baldwin and Robichaux, 1995) of Hawaii.

The Hawaiian *Tetramolopium* lineage has *T. humile* as its basal species (Lowrey, 1995). This species occurs only in alpine and subalpine habitats on Hawaii and Maui, and occupies an ecological niche equivalent to its putative ancestor, which is found in New Guinea. *Tetramolopium humile* most likely colonized one of those islands, then migrated to the other. The genus spread into xeric, open lowlands and submesic dry forests on Hawaii and Maui and other islands. One species from Molokai even reached the Cook Islands, where it diverged from its Hawaiian progenitor.

The invasion of "new islands" or "new habitats" may be accompanied by speciation, especially in the case of the latter. As observed by Carlquist (1974), "Speciation . . . progresses by means of continual shifts into new habitats and habits." The relative incidence of intra-island and inter-island speciation depends on the level of environmental heterogeneity within islands and the distance between islands. In *Schiedea* and *Alsinidendron* (Caryophyllaceae), which comprise one of the largest endemic radiations among Hawaiian angiosperms, 14 species appear to have originated following inter-island colonization and 42 following intra-island niche shifts (figure 2.8, Wagner et al., 1995).

The *Schiedea globosa* clade ostensibly originated on Oahu. There were 18 hypothesized colonizations from older to younger islands, 11 of which were followed by speciation. There were also four hypothesized back-colonizations from younger to older islands, three of which were followed by speciation.

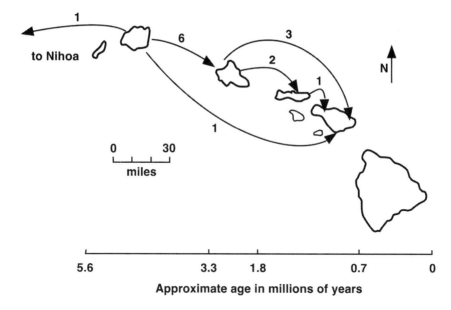

Figure 2.8. Summary of colonization events associated with speciation in *Schiedea* and *Alsinidendron* (from Wagner et al., 1995. Redrawn with permission of the author).

In *Argyranthemum* and several other genera of the Canary Islands, nearly all speciation is associated with inter-island colonization (Francisco-Ortega et al., 1997). In *Argyranthemum* there were two major radiations. One is in areas affected by the northeast trade winds including humid lowland scrub and laurel forest. The other is in regions less affected by the winds, including coastal and high-altitude desert and pine forest. In general the progression is from older to younger islands, suggesting that niches get filled and competitive exclusion predominates. This is the case in the aforementioned genera of the Hawaiian Islands.

Adaptive radiation often is relatively rapid as suggested by the youth of the islands, the high degree of certainty of area cladograms, the cross-fertility of some species, and the contrast between moderate phenotypic divergence and little allozyme divergence (Crawford, Stuessy, and Silva, 1987; Carlquist, 1995). Consider, for example, the Hawaiian *Tetramolopium* (Lowrey, 1995). This monophyletic assemblage, based on a New Guinea antecedent, has evolved into three clades that are ecologically quite distinctive. The genetic identity of the most divergent species is 0.87, which is within the range of conspecific populations in continental areas. From this value, the time of their divergence is estimated at 600,000 to 700,000 years before present.

While rapid radiations are best known on islands, they also occur on large land masses. In the genus *Aquilegia*, the rapid radiation of 14 species is deduced from very low levels of sequence divergence in both a nuclear DNA region (the ITS region of rDNA) and chloroplast DNA region (the space between the *atp*B and *rbc*L genes; Hodges and Arnold, 1994). The species also are largely interfertile. This radiation followed the evolution of floral nectar spurs. Interspecific differences in nectar spur morphology adapt species to different pollinators and thus represent a form of ecological divergence. There are multiple, independent origins of floral nectar spurs throughout the angiosperms, and they appear to be correlated with increased rates of species diversification (Hodges, 1997).

Continents also offer opportunities for radiation when a new set of habitats is made available, as with the retreat of glaciers. The recent glaciation in the northern Andes has created a new set of conditions associated with low temperature, high humidity, drought, and nutrient shortage. *Espeletia* (Asteraceae) has colonized this area and radiated into 63 species varying in habit from giant rosettes to trees, and in stem morphology, energy allocation, and leaf pubescence (Monasterio and Sarmiento, 1991). The same trends in several traits with altitude are also present in the 11 species of the giant senecios *Dendrosenecio* (Asteraceae), which are endemic to the tall mountains of eastern equatorial Africa. Chloroplast restriction-site analysis shows that this radiation was very recent and occurred within mountains as well as between them (Knox and Palmer, 1995).

Genetic Constraints to Radiation

The rapid diversification of lineages on islands has occurred in spite of the genetic bottlenecks that ostensibly accompanied the initial and subsequent colonization events (Carlquist, 1974). Bottlenecks may be associated with a loss of genetic variation and considerable inbreeding if populations are quite small. The reduction in level of heterozygosity depends on the rate of population growth as well as on the size of the bottleneck (Nei, Maruyama, and Chakraborty, 1975). If population growth is rapid, the reduction in heterozygosity is small, even if the number of colonists is small. Conversely, the average number of alleles lost per locus is quite sensitive to bottleneck size, but not the rate of subsequent population growth. Protracted periods of small population size following colonization are the most detrimental to the retention of genetic variation.

The level of inbreeding and amount of genetic variation following a colonization event depend on the number and source of colonists, and on the relatedness of colonists. The fewer the number of colonists, the closer their relatedness; and the fewer the number of source populations, the greater the level of inbreeding and the lower will be the amount of genetic variation (Whitlock and McCauley, 1990). The extreme case would be where all colonists are closely related. This has been referred to as kin-structured migration (Fix, 1978). In plants one fruit may be associated with a given colonization event. Outcrossed seed in that fruit would be full sibs if they shared a sire, or half sibs if they did not. The coefficient of relationship between half sibs is 0.25 versus 0.50 for full sibs if the sires were unrelated, and 0.67 if the seeds were the products of selfing (Squillace, 1974).

The effect of kin-structured migration on nuclear genetic variation is different in degree than its effect on chloroplast or mitochondrial DNA variation. In the most extreme case, migration is by propagules produced asexually by the same plant. There would be neither cytoplasmic nor nuclear variation. If colonization is by seeds from a single fruit, there would be no cytoplasmic variation but some level of nuclear variation, assuming maternal inheritance of cytoplasmic markers. After spread from the site of colonization, there would be much greater nuclear as opposed to cytoplasmic variation among populations. Eventually cytoplasmic heterogeneity would increase due to mutations.

Wade, McKnight, and Shaffer (1994) have constructed a model for estimating time of colonization of an isolated area based on the disparity between nuclear and cytoplasmic variation. This model would be very useful in estimating when a genus arrived on a distant island. Such an arrival time has been estimated from the age of islands and from the allozyme genetic identity of or DNA mutational steps between the most divergent species (Wagner and Funk, 1995).

The loss of genetic variation accompanying long-distance colonization is well documented in *Eichhornia paniculata* (Pontederiaceae). The species, which has three style morphs (short, mid, long), colonized Jamaica from a South American source, most likely Brazil. The introduced populations are polymorphic at 8% of the loci measured vs. 24% in the source populations; and the former has a mean heterozygosity of 0.02 vs. 0.08 in the latter (Glover and Barrett, 1987). None of the populations on Jamaica has the short-style morph, and thus are missing the S allele at one of the two loci controlling style length (Barrett, Morgan, and Husband, 1989). The absence of this allele in Jamaica probably is due to its loss during or subsequent to colonization.

Another example of the loss of genetic variation accompanying colonization is in the pitcher plant *Sarracenia purpurea*. This case is especially interesting because we know the number of founders and time of establishment. The Cranberry Island population was founded in 1912 by a single plant, which was introduced by a graduate student. Schwaegerle and Schaal (1979) analyzed 10 allozyme loci and found that the aforementioned population has only one polymorphic locus versus 2.5 for 11 typical populations. Mean heterozygosity per individual across loci is 0.042 at Cranberry Island versus 0.089 for all populations.

The number of colonists and colonization episodes, the sources of colonists, and their breeding system have a major impact on variation in the introduced populations relative to the source populations. Comparisons have been made between such populations in several species. Contrary to expectation, introduced populations do not invariably have less variation than the populations from which they were derived, as seen in *Echium plantagineum* (Boraginaceae) and the grass *Apera spica-venti* (Brown and Marshall, 1981; Barrett and Husband, 1990). Both species are native to Europe, and were introduced into Australia and Canada, respectively.

The Mobilization of Genetic Variation

How does radiation occur in an assemblage that is initially genetically depauperate? There are three ways to enhance the number of alleles subject to selection: mutation, the breakdown of developmental canalization, and hybridization. Mutation can gradually increase genetic variation, especially if mutation rates were heightened by inbreeding or environmental stress (McClintock, 1984; McDonald, 1995). Inbreeding and external stress may reduce the precision of developmental canalization and expose underlying genetic variation (Levin, 1970).

Given that the breakdown of developmental canalization rarely is discussed, some examples are in order. Consider first the effect of stress. In *Linanthus* (Polemoniaceae), corollas normally have five corolla lobes, but occasionally a corolla has four or six to seven lobes (Huether, 1968). Entire plants do not deviate from the pentamerous condition. Under stressful greenhouse conditions the incidence of abnormal corolla lobes increased. Moreover, this abnormality could be captured by artificial selection. After five generations of selection for abnormally high corolla lobes, about 55% of the flowers in the selected line had more than five lobes vs. 8% on the control line.

Meristic instability also is affected by inbreeding. Lehmann (1987) grew two wild populations and eight cultivars of *Phlox drummondii* (Polemoniaceae) in the same greenhouse. Whereas only 2% of the flowers on wild *Phlox* deviated from five petals, between 4 and 27% of the cultivar flowers (depending on the cultivar) had other than five petals. Most deviant flowers had six petals. Deviations in sepal, anther, and carpel numbers also were elevated in the cultivars. Aberrant numbers in one whorl are sometimes accompanied by aberrations in other whorls, leading to flowers with a plan of 6 sepals-6 petals-6 stamens-4 carpels. The cultivars are autogamous, whereas the wild plants rarely are so (Levin, 1976b). The former have about half the mean heterozygosity of their wild relatives.

Once radiation has commenced, hybridization between close relatives can enrich local gene pools and set the stage for subsequent evolution (Carlquist, 1974). The role of hybridization in adaptive radiation is suggested by the intermediate character expression of some species and by the incongruence between chloroplast DNA and nuclear DNA trees. Hybridization may lead to diversification within species as well as the origin of new species. For example, it is likely that in the Canary Islands, subspecies of *Argyranthemum adauctum* occur in laurel forests and heath belts because of the incorporation of genes from related species indigenous to those habitats (Francisco-Ortega et al., 1997). Hybridization is thought to have been important in the radiation of species in *Bidens, Gouldia, Pipturus,* and *Scaevola* (Gillett, 1972), *Cyrtandra* (Smith, Burke, and Wagner, 1996), and the silversword alliance (Baldwin, 1997) in Hawaii.

Hybridization also may destabilize the genetic environment as seen in another member of the Polemoniaceae. *Gilia millefoliata* × *G. achilleaefolia* F_1 and F_2 hybrids display normal floral forms whereas many F_3 and F_4 hybrids exhibit 4-,7-,8-, and 9-merous flowers (Grant, 1956). The aberrant plants fail to breed true, with progeny being normal or having different aberrant merisms that vary within and between plants. An allotetraploid of the aforementioned hybrid also displays meristic abnormalities.

Is developmental instability of any evolutionary significance? Can the meristic novelties be stabilized? If the expressions released through developmental instability have adaptive value, they may be favored by selection, their penetrance and expressivity enhanced, and the expressions stabilized. Apparently this happened in *Leptodactylon jaegeri* (Polemoniaceae), which has a floral plan of 6 sepals-6 petals-6 stamens-4 carpels. Other members of the genus have a floral plan of 5-5-5-3, as do nearly all members of the Polemoniaceae (Grant, 1956).

The corollas of *Dodecatheon* species consistently are 5-merous, including those of *D. hendersonii* and *D. clevelandii*. However, *D. hendersonii* subsp. *cruciatum*, which may be a product of hybridization between these species, has 4-merous corollas (Stebbins, 1969b). *Ceanothus jepsonii*, also believed to be a hybrid derivative, has six or seven sepals, petals, and stamens, whereas no other species in the Rhamnaceae has such numbers (Anderson and Stebbins, 1954).

Nonadaptive Radiation

The passage through bottlenecks enhances the potential for nonadaptive divergence. Nonadaptive radiation may accompany adaptive radiation as independent evolutionary transitions or nonadaptive radiation may arise through the genetic correlations of unselected traits and selected traits. Selection on one character expression may, through pleiotropy or linkage, result in changes in other characters—that is, in correlated responses to selection. Some such responses may have a negative effect on the population and thus retard the evolution of the character under selection. In an open environment or one with relaxed selection pressures, as is envisioned for oceanic islands, nonadaptive changes may be tolerated, whereas they may not be on continents (Carlquist, 1974; Crawford et al., 1987).

The most suggestive evidence for nonadaptive radiation comes from within and between species inhabiting the Aegean Archipelago (e.g., *Nigella*, Ranunculaceae; Strid, 1970; *Erysimum*, Cruciferae; Snogerup, 1967). Small populations differ in single traits or combinations of traits that apparently are haphazard in kind and in space. The habitats occupied by many species vary little among islands and the taxa are ecologically unspecialized.

On Naxos, plants of *Nigella degenii* ssp. *degenii* have pale blue anthers and pale yellow pollen, whereas on the neighboring island of Mikonos plants of *N. degenii* ssp. *barbro* have red anthers that contain either violet or yellow pollen (Strid, 1970). Pollen color varies markedly among populations, from 2% violet and 98% yellow in one population to 76% violet and 24% yellow in another (figure 2.9). In *Erysimum* conspecific populations occupying similar habitats on different islands may differ in the frequency of petal claws, petal shape, and the presence or absence of leaf serrations.

Striking differences also occur among populations of *Pedicularis dasyantha* on the island of Svalbard (north of Norway). Odasz and Savolainen (1996) found seven distinctive corolla pigmentation morphs varying from dark purple to light purple and light pink. Each of these morphs is fixed in at least one of the 13 populations studied. Some populations have as many as five morphs. The dark morph coincides with allele 1 at the *6-pgd* locus, and the light morphs with allele 2. The species is self-compatible, and apparently is free of strong

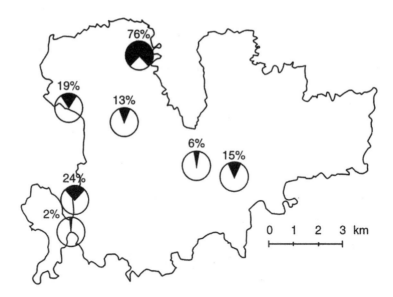

Figure 2.9. Variation in violet pollen in seven populations of *Nigella degenii* on Mikonos (from Strid, 1970, Fig. 27. Redrawn with permission of the Council for Nordic Publications in Botany).

stabilizing selection by pollinators. The only potential pollinators observed on Svalbard were species of Diptera and *Collembola*. Closely related species growing elsewhere in the arctic are pollinated by *Bombus* species, which do have corolla color preferences.

Carlquist (1974) suggests that some of the differences separating the 126 species of *Cyrtandra* (Gesneriaceae) on Oahu (Hawaiian Islands) may be non-adaptive. Similarly, Morley (1972) observes that some interspecific differences between species of *Columnea* (Gesneriaceae) in western Jamaica seem to lack adaptive significance. The extent to which these patterns of variation actually are due to stochastic processes remains to be determined. Differences in chromosomal arrangements between the species, and a paucity of genetic variation within species but conspicuous differences between them, point to an important role for stochastic processes in their histories.

The passage of populations through bottlenecks also enhances the possibility of chromosomal evolution. Given chromosomal polymorphism, the smaller the effective population size the greater is the probability of fixing a novel arrangement, even if heterozygotes have reduced fertility (Lande, 1985). Chromosomal rearrangements are a prerequisite for aneuploid changes in chromosome number.

Assuming that bottlenecks were part of the demographic and evolutionary history of most plant species that evolved on oceanic islands, it is surprising they display little evidence of chromosomal evolution. Only in the silversword alliance do we see reciprocal translocations and changes in chromosome number (Carr and Kyhos, 1981, 1986).

Overview

Species are composed of populations of various sizes and with various levels of genetic variation. Quantitative genetic studies and selection programs indicate that large populations of many species have considerable potential to alter their ecological strategies. How often populations respond to selection is another matter. We better appreciate the potential for change than the constraints on change. Because there are so many populations subjected to disparate selection pressures over long periods of time, only a very small percentage of populations need respond in a meaningful way to generate a large amount of diversity.

We would like to gain some insights on the tempo of evolution. Unfortunately, this parameter eludes us. The rate of single-character evolution in natural populations is largely unknown, except in introduced species where we have a known time frame. Many characters are functionally related. How rapidly these character complexes evolve is even harder to estimate. Character complexes have shifted in the same direction more than once in many genera. The rate of evolution could have been quite different across lineages.

Ecological shifts apparently proceed most rapidly when hybridization and/ or polyploidy are involved. Hybridization provides a burst of variation on which selection may act, whereas polyploidy affords novelty through the effect of chromosome doubling per se, as well as through the simple coupling of different genomes.

The progression of ecological shifts is best understood on oceanic islands, where relatively low levels of competition allow phylads to capitalize on an abundance of ecological opportunities. We can understand the temporal context in which diversification occurred from molecular phylogenies and from the ages of the islands themselves. We also can see that ecological radiation does not of necessity bring with it pronounced genomic discord.

As variation is a prerequisite for radiation, we may ask where the variation comes from. Mutation is an obvious source, but not one to release variation very quickly. Alternative sources, such as the breakdown of developmental canalization, have hardly been considered.

3

Speciation
The Genetic Transition

The genetic changes leading to the formation of new species are poorly under-stood and are the subject of considerable debate. Some evolutionists propose that major gene changes may propel populations into different reproductive or ecological realms and thereby promote speciation (Gottlieb, 1984). Specia-tion may be a direct result of selection acting on these genes or a pleiotropic consequence of adaptive shifts elsewhere in the genome (Templeton, 1980). Others propose that the interactive effects of many minor gene loci produce divergent reproductive or ecological profiles (Barton and Charlesworth, 1984; Coyne and Lande, 1985). These views are not mutually exclusive; some differ-ences may be dictated by major genes, while others may be dictated by poly-genes.

I will not evaluate the merits of each of the views. Indeed all of the afore-mentioned genetic changes are involved to various degrees in speciation. Rather, the genetics of reproductive and ecological discontinuities will be con-sidered on a character by character basis, recognizing that species will differ by more than one character. The genetics of genomic incompatibility, including cross-incompatibility, and hybrid weakness and sterility also will be considered.

Genetic Control of Interspecific Differences

The genetic basis of differences between species that confer pre- and postzy-gotic barriers to gene exchange has been difficult to determine because crosses often are incompatible or hybrids are inviable or sterile. Because the strength of pre- and postzygotic barriers are positively correlated with the degree of genetic divergence (Levin, 1978a), the genetic basis for differences differentiat-ing species will be best understood for closely related species. If we analyze species that are distantly related, we may overestimate the number of genes initially involved in reproductive isolation, because most of the change may have followed the speciation event (Coyne, 1992). Indeed most of the following examples do not involve recent progenitor-derivative pairs, so their value in understanding speciation genetics is somewhat reduced. Nevertheless, the forthcoming estimates are of value because the number of genes controlling a trait, the magnitude of their effects, and their position in the genome affect the speed and mode of divergence (Fisher, 1958; Mitchell-Olds and Rutledge, 1986).

Floral Characters

Major genes and minor genes can both be responsible for interspecific differences in floral attributes. The classic study on the floral morphology of *Aquilegia* by Prazmo (1965) illustrates the effects of major genes. Based on F_2 segregation patterns, he found that one locus determines whether spurs are straight or curved, a trait that differs among species groups. A second locus controls whether the flower is erect or nodding, the latter being dominant. Erect flowers are visited by lepidopterans, whereas nodding flowers are visited by bees and hummingbirds. A pair of duplicate loci govern whether the flower has a spur on each of its five petals. Spurs are the dominant condition.

Species of the Asteraceae sometimes differ by the presence/absence of ray florets as in *Layia discoidea* and *L. glandulosa*. The former is discoid and a likely derivative of the radiate *L. glandulosa*. Ray presence/absence is governed by two independently assorting loci (Clausen, Keck and Hiesey, 1947; Ford and Gottlieb, 1990). In *Haplopappus* the presence/absence of ray florets is under single-gene control (Jackson and Dimas, 1981). This difference may impose a partial ethological barrier between the species, because pollinators are attracted more to radiate variants than to discoid ones in *Senecio* (Abbott and Irwin, 1988) and *Bidens* (Sun and Ganders, 1990).

Bradshaw, Wilbert, and Schemske (1995) demonstrated that major genes are involved in the floral divergence and isolation of the bee-pollinated *Mimulus lewisii* (Scrophulariaceae) and the hummingbird-pollinated *M. cardinalis*. Flowers of *M. cardinalis* have red petals lacking nectar guides, a tubular corolla, a large volume of nectar, and exerted anthers and stigmas. Flowers of *M. lewisii* have pink petals with yellow nectar guides, a wide corolla orifice, a small volume of highly concentrated nectar, and inserted anthers and stigmas. Using DNA polymorphisms and genome mapping techniques, Bradshaw et al. (1995) demonstrated that each difference in petal anthocyanins and carotenoids, corolla width, petal width, nectar volume, nectar concentration, and stamen and pistil length is determined by at least one gene of major effect—that is, a gene accounting for more than 25% of the phenotypic variance among F_2 plants. This study and conventional analyses of F_2 segregation patterns show that the petal color difference is controlled by one gene and that other differences are controlled by several genes (Hiesey, Nobs, and Bjorkman, 1971).

Studies on *Mimulus* also demonstrate that some floral traits are under complex genetic control. *Mimulus cupriphilus*, a predominantly selfing species, ostensibly is a recent derivative of the mixed mating *M. guttatus*. The former has smaller flowers and no petal spots. Based on estimates of minimum gene number from F_2 segregation patterns, at least three to seven genes control differences in flower size, and at least three genes control flower spot number (Macnair and Cumbes, 1989). Differences in corolla lobe shape are controlled by one major gene.

Another recent derivative of *M. guttatus, M. micranthus*, also is predominantly selfing. The evolution of small flower size in the latter is associated with more rapid floral growth and shorter duration of floral development (Fenster et al., 1995). Segregation patterns in the F_2 generation indicate that the species differ in 8 and 10 factors for the growth rate and duration of bud development, respectively. The selfer is characterized by smaller corolla width and reduced pollen production. The minimum number of loci differentiating the species for these traits are 8.7 and 13.5, respectively (Fenster and Carr, 1997).

Geum coccineum and *G. rivale* produce hybrids that are fully fertile and exhibit normal chromosome pairing (Gajewski, 1950). In *G. coccineum* the sepals are reflexed and the petals are horizontal. In *G. rivale* the sepals and petals are erect. The F_2 segregation pattern was skewed toward the former, but nevertheless suggested polygenic control for floral orientation. The petals of *G. rivale* are 6–10 mm wide, while those of *G. coccineum* are 12–27 mm wide. The binomial regularity of the F_2 progeny for petal width is indicative of polygenic control. The difference between the cream-colored petals of *G. rivale* and the orange-red petals of *G. coccineum* is due to two pairs of complementary genes and perhaps additional modifiers.

Finally, the genetic basis for interspecific differences in some quantitative floral traits have been studied in *Nicotiana*. Differences in corolla size and shape (East, 1916), and differences in the ratio of corolla limb to tube length in the bud and their relative rate of elongation (Smith, 1950) are polygenic.

Ecological Characters

The genetic architecture of ecological differences between related species is poorly understood. Presumably some combination of major and minor genes are involved. Major genes have been implicated in heavy metal tolerance in several species (Macnair, 1993). The difference in habitat preference of *Mimulus guttatus* and its copper-tolerant derivative *M. cupriphilus* ostensibly is controlled by a major gene and modifiers (Macnair and Cumbes, 1989).

We may look at the basis for alternate edaphic preferences within species for clues to explain differences between species. Consider *Silene vulgaris*. Two distinct major genes are responsible for zinc tolerance (Schat et al., 1996), and two major additive genes also appear to control copper tolerance in this species (Schat et al., 1993).

The genetic basis of alternative flowering time has been studied within *Arabidopsis thaliana*. Kuittinen, Silanpaa, and Savolainen (1997) mapped quantitative trait loci (QTLs) for flowering time in the progeny of a cross between an early-summer and an overwintering Finnish population. They found that the trait is controlled by one major and six minor QTLs, the former explaining 53% of the variance in flowering time.

Change from the perennial to annual habitat has accompanied speciation in many genera, but the genetic basis for the change is poorly understood. Some insights are available in *Mimulus*. Macnair and Cumbes (1989) crossed the perennial *M. guttatus* and derivative annual *M. cupriphilus*. The F_1 hybrids are perennial, thus indicating dominance for this expression. The annual habit segregates in advanced generation and backcross hybrids. The number of genes involved could not be determined.

Many close relatives have divergent growth forms (e.g., prostrate vs. erect) because of differences in apical dominance. The domestication of crop plants often has involved an increase in apical dominance, and here we have some insights into its genetic control. For example, the profound increase in the apical dominance of maize relative to its probable wild ancestor, teosinte, is due largely to a single gene (*teosinte branched 1*). The maize allele of *tb1* is expressed at twice the level of the teosinte allele (Doebley and Wang, 1997).

There is considerable interest in the genetic control of rhizomatousness, as this trait has a bearing on the weediness of species. One weed that spreads very effectively by rhizomes is johnsongrass, *Sorghum halepense*, an allotetraploid. The control of rhizome production was determined in the progeny of crosses between its putative diploid parents, *S. bicolor* and *S. propinquum*, by Paterson, Schertz et al. (1995). They found that this trait is influenced by eight QTLs; gene action in four is additive, in two is dominant, in one is recessive, and in one is overdominant.

The mapping of quantitative and other traits onto chromosomes (or linkage groups) reveals surprising similarity in the genetic architecture of shared traits among members of the same taxonomic families. For example, a close correspondence among QTLs affecting seed size, disarticulation of the inflorescence, and day-neutral flowering time has been shown for rice, barley, wheat, sugarcane, sorghum, and maize (Paterson et al., 1995). A correspondence among a high proportion of genes affecting height has been shown for maize and sorghum (Lin, Schertz, and Paterson, 1995).

As more information is forthcoming, we will see whether QTL analysis in one taxon typically has much predictive value for other taxa within different genera of the same family. Extensive conservation of gene repertoire and order seems to be the case in cultivated cereals (Paterson, 1995), and in several genera including *Brassica* (Lagercrantz and Lydiate, 1996), *Helianthus* (Rieseberg et al., 1995), and *Solanum* (Bonierdale, Plaisted, and Tanskley, 1988).

Given the conservation of gene repertoire and to a large extent gene order, it is possible that parallel changes associated with speciation in different parts of a genus are dictated by changes at the same loci. For example, it is possible that the multiple and independent shifts from a perennial to an annual habit in *Gilia* are controlled by the same loci. Support for parallel changes at corresponding genetic loci comes from studies on the genetic control of large seeds, nonshattering seeds, and day-length insensitive flowering in sorghum, rice, and maize (Paterson, Lin et al., 1995). These similar phenotypes are largely determined by a small number of QTLs that correspond closely in location in the three species. Correspondence in location does not prove that the genes are identical, but it suggests that some of them may be. This idea is bolstered by the tendency of corresponding QTLs to have similar gene action (Paterson, Lin et al., 1995).

Cross-Incompatibility

Interspecific cross-incompatibility is a complex process involving the timing and level of pollen grain germination on heterospecific stigmas, and the rate and length of tube growth and its ability to penetrate the micropyle. Interspecific cross-incompatibility may be controlled in part by the same locus (S) that controls self-incompatibility. Alterations in S-alleles and in S-allele expression by modifier genes have been correlated with shifts in interspecific cross-compatibility in *Nicotiana* and *Lycopersicon* (de Nettancourt, 1977). In *Lycopersicon* the major control of unilateral incompatibility between *L. esculentum* and *L. hirsutum* (crosses are successful only when the former was the egg parent) can be attributed to the S locus, but two additional loci on other chromosomes also affect crossability (Bernacchi and Tanksley, 1997). Simple genetic control by other loci also has been reported. Two genes (Kr_1 and Kr_2) have been implicated in the partial crossing barrier between *Triticum* and *Secale* (Jalani and Moss, 1980). A single dominant gene conditions partial incompatibility between *Hordeum vulgare* and *H. bulbosum* (Pickering, 1983). In contrast cross-incompatibility among conspecific populations of *Phacelia dubia* (Hydrophyllaceae) is polygenic (Levy, 1991). Which of these genetic architectures is the norm remains to be determined.

Hybrid Weakness and Lethality

The lethality or weakness of some hybrids is due to the complementary action of alleles at one or a few loci. Perhaps the best-known example of an aberrant developmental pattern in hybrids caused by a single gene is the "corky" syndrome expressed in some hybrids between *Gossypium hirsutum* and *G. barbadense* (Stephens, 1950). Young stems and occasionally leaf petioles and mid-ribs develop a corky surface. The syndrome also includes the suppression of apical dominance, profuse branching, and dwarfism. The "corky" syndrome is based on the genotype $ck^x ck^y$; the former allele comes from G. *hirsutum* and the latter comes from G. *barbadense*.

First-generation hybrids between *Oryza breviligulatus* and *O. glaberrima* often are weak. This condition is controlled by two complementary dominant genes, W_1 and W_2, that are common in O. *breviligulatus* and O. *glaberrima*, respectively (Chu and Oka, 1972). Two complementary dominant lethals are responsible for chlorosis in hybrids between varieties of O. *sativa* (Oka, 1957). Complementary genes causing hybrid inviability also occur in *Crepis* (Hollingshead, 1930) and *Gossypium* (Gerstel, 1954).

In many hybrid combinations, the F_1 plants are vigorous and fertile, but segregates in the F_2 are either weak or partially sterile (Stebbins, 1958; Grant, 1981). Advanced-generation breakdown usually has been attributed to genic imbalance across numerous loci. However, breakdown may have a simple genetic basis. For example, in *Setaria italica* where some F_2 hybrids have poor growth, the segregation of normal to weak was 15 : 1 (Croullebois et al., 1989). This indicates that weakness may be due to two complementary recessive genes. Chlorosis in F_2 hybrids of crosses between some *Melilotus* species is based on two sets of duplicate recessive genes (Sano and Kita, 1978).

Two complementary recessive genes also are responsible for a reduction in vigor in F_2 hybrids derived from different strains of *Oryza sativa*. Of particular interest is their approximate location in the genome. Fukuoka, Namai, and Okuno (1998) mapped each gene independently in F_2 populations derived from two strains. One gene (*hwd 1*) was mapped onto the distal region of chromosome 10, and the other gene (*hwd 2*) was mapped onto the central region of chromosome 7.

Crosses between many combinations of species yield various levels of seed abortion (Levin, 1978a). Seed abortion is due to disharmonious interaction among the three types of tissue in the developing seed—that is, maternal tissues, embryo, and endosperm. The genetic basis for hybrid seed abortion is poorly understood. One informative pair of studies was conducted on *Lens*. Ladizinsky, Cohen, and Muehlbauer (1985) obtained an interspecific hybrid between *L. ervoides* and *L. culinaris* using embryo rescue techniques. This plant produces half normal and half aborted embryos upon selfing. Pod abortion ranges from 10 to 90% in the F_2 progeny, implying quantitative control. A later study indicated that the abortion of F_1 hybrids is strongly affected by dominant or epistatic gene interaction (Abbo and Ladizinsky, 1994).

There is no evidence that the aforementioned genes involved in hybrid weakness or inviability have any adaptive significance. Thus it is of interest to consider the basis for inviability in some hybrids between a copper-tolerant species, *Mimulus cupriphilus*, and its nontolerant progenitor, *M. guttatus* (Macnair and Christie, 1983). Two independent genetic systems afford partial reproductive isolation. One gene that confers copper tolerance, or another tightly linked to it, interacts with a small number of genes in *M. guttatus* to give seedling death in hybrids in the fourth-leaf stage or later. Another gene that confers copper tolerance, or another tightly linked to it, produces hybrid seedling death before the first true leaves expand.

Hybrid Sterility

The sterility of hybrids may be due to the interaction of cytoplasmic and nuclear factors in gametes, the recombination of genes belonging to disparate genomes during meiosis, abnormal pairing of chromosomes during meiosis, or the failure of reproductive organs to form in a normal manner. Given that this discussion is concerned with the genetic bases for interspecific character differences and genome incompatibility, the nuclear-cytoplasmic basis for hybrid sterility will be considered first. Discord between nuclear and cytoplasmic genes is manifested in reciprocal crosses. For example, in *Oryza sativa* (rice) hybrids between subspecies *japonica* and subspecies *indica* spikelet sterility is 17% when the former is the female parent versus 51% when the latter is the female parent (Li et al., 1997). In hybrids with ssp. *japonica* as the female parent, pollen sterility is 38% versus 57% when *indica* is the female parent.

In some instances, reciprocal-specific effects are associated only with male function. For example, hybrids between *Phacelia dubia* var. *dubia* and *P. dubia* var. 'railroad' have an average of 90% pollen fertility when the former is the egg parent versus 65% fertility when the latter is the egg parent (Levy, 1991).

When male function varies as a reciprocal-specific effect, crosses in one direction often yield hybrids that are male sterile, whereas crosses in the other direction yield hybrids with moderate to normal male function (Kaul, 1988). Typically one or two nuclear genes are involved in male sterility. For example, male sterility in hybrids between *Nicotiana langsdorffi* and *N. sanderae* is due to two genes, each of which occurs in two allelic forms (Smith, 1968). In hybrids between *Helianthus annuus* and *H. grosseserratus*, male sterility is conditioned by one gene whose expression is environment-dependent (Kaul, 1988).

Recently interest has turned to the location in the cytoplasm of the elements causing male sterility. In *Epilobium* there is altered mitochondrial gene expression in male sterile hybrids, which suggests that mitochondria are involved with male sterility (Schmitz and Michaelis, 1988). Chloroplast gene expression is unaltered in these hybrids (Schmitz, 1988). There are insufficient data to say whether mitochondrial gene action is associated with male sterility in hybrids of other genera.

In most plant hybrids, sterility is associated with differences in chromosomal homology or genic disharmony in the haploid generation. Some hybrids have normal meiosis, but are sterile nevertheless (cf. Levin, 1978a). Many hybrids have abnormal chromosome pairing. This can be due to the action of specific genes that affect the stringency of chromosome pairing (Jenkins and Jimenez, 1995), differences between species in the amount, nature and dispersion of repetitive DNA (Jones and Rees, 1968) and interspecific differences in chromosomal arrangements. Stebbins (1958) contends that the latter is the decisive factor in determining the sterility of F_1 hybrids.

The chromosomal basis for reduced hybrid fertility has been described in hybrids between several pairs of progenitors and neospecies. For example, *Clarkia lingulata* differs from its putative progenitor *C. biloba* by a translocation and a paracentric inversion (Lewis and Roberts, 1956). The former has a haploid number of 9 versus 8 in the latter. Hybrids have less than 15% viable pollen.

Chaenactis glabriuscula (Asteraceae) ostensibly has given rise to *C. stevioides* and *C. fremontii* (Kyhos, 1965). The parental species has a haploid chromosome number of 6 versus 5 in both derivatives. The derivatives differ from the parental species by at least two translocations (figure 3.1). These translocations are not shared by the derivatives as seen by abnormal chromosome pairing in their hybrids. Indeed *C. stevioides* and *C. fremontii* differ by two translocations. The fertility of hybrids between *C. glabriuscula* and each of its derivatives varies from about 10 to 50%.

Chromosomal differences need not be large to cause a reduction in fertility. Stebbins (1958) notes that many sterile hybrids with normal or nearly normal chromosome pairing were rendered fertile by artificial doubling of chro-

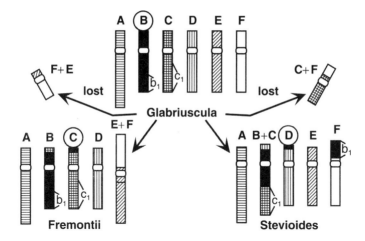

Figure 3.1. Probable structural arrangement of the chromosomes of *Chaenactis glabriuscula* and its derivatives (from Kyhos, 1965. Redrawn with permission of *Evolution*).

mosome number. Had sterility been under genic control, it would have persisted in the doubled hybrid.

In general, fertility declines as the number of chromosomal differences between species increases. Large rearrangements have a greater impact on fertility than very small ones. Chandler, Jan, and Beard (1986) sought the relationship between the number of chromosomal differences and pollen fertility in 34 types of interspecific hybrids (F_1s) in *Helianthus*. Hybrids heterozygous for three or more translocations have lower fertility on average than those with none or few (figure 3.2). The correlation between the number of translocations and

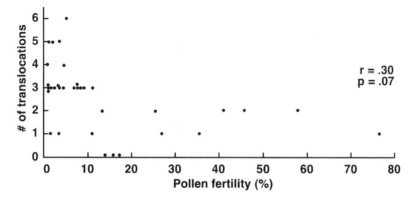

Figure 3.2. The relationship between pollen fertility in hybrids and the number of translocations by which *Helianthus* species differ (after Chandler et al., 1986).

pollen fertility is $r = 0.30$, which falls short of statistical significance. Pollen fertility in *Helianthus* also is diminished by inversion heterozygosity and the occasional failure of chromosomes to pair.

The mapping of quantitative trait loci in intersubspecific rice hybrids has uncovered supergenes that may be "cryptic structural rearrangements," specifically inversions, on two chromosomes (Li et al., 1997). Supergenes are groups of tightly linked genes within which recombination will cause reduced fitness (Darlington and Mather, 1949). Pollen sterility is associated with heterozygotes and ostensibly occurs as a result of recombination within the supergenes. The mapping of 117 loci in hybrids between *Tetramolopium humile* and *T. rockii* (Asteraceae) reveals a high degree of segregation distortion, which may be due to cryptic structural hybridity (Whitkus, 1998). The hybrids show no reduction in fertility.

The nature and magnitude of chromosomal rearrangements that contribute to hybrid sterility and by which species are differentiated have been elusive because of a paucity of chromosomal markers. The utilization of a broad range of genetic markers now permits the rather dense genetic mapping of plant chromosomes. From such maps one can discern chromosomal rearrangements of various sizes. Most comparative chromosome mapping has involved species of agronomic importance (e.g., *Lycopersicon* and *Capsicum*; Tanksley et al., 1988; *Sorghum* and *Zea*; Berhan et al., 1993; *Triticum* and *Hordeum*; Dubcovsky et al., 1996). Such mapping reveals that species in different genera may differ by several translocations and paracentric inversions. Even divergent species within a genus may differ by several rearrangements (e.g., *Brassica*; Lagercrantz and Lyiate, 1996).

One of the first applications of chromosome mapping in the study of hybrid sterility involved *Lens culinaris* and *L. ervoides*. The species were thought to differ by a single translocation. Using a segregating F_2 population, Tadmor, Zamir, and Ladizinsky (1987) were able to correlate four isozyme markers with quadrivalent formation during meiosis and identify the translocation end points. Plants with pollen viability below 65% are heterozygous for the translocation, while plants with viability above 85% are homozygous, thus implicating the translocation as the primary basis for reduced hybrid fertility.

A study of the same type was conducted in *Helianthus*. *Helianthus annuus* and its close ally *H. argophyllus* differ by two reciprocal translocations. Quillet et al. (1995) analyzed the segregation of isozyme, RAPD and morphological markers in backcross (BC_1) progeny. Over 80% of the variation in pollen fertility is explained by genetic intervals located on three linkage groups. Meiotic abnormalities are tightly correlated with the markers circumscribing these intervals, indicating that the translocations were the prime cause of reduced hybrid fertility.

Of special interest here is the chromosomal architecture of species and their derivatives. The most informative data set is on *Helianthus*, and involves *H. annuus*, *H. petiolaris*, and their hybrid derivative, *H. anomalus*. Rieseberg et al. (1995) generated linkage maps based on 212 loci in *H. annuus* and 400 loci in *H. petiolaris*. The genetic markers were mapped to 17 linkage groups, corresponding to the haploid chromosome number of the three species. By comparing the genomic location and linear order of homologous markers, Rieseberg et al. were able to infer chromosomal structural relationships among the three species (figure 3.3).

Helianthus annuus and *H. petiolaris* differ by at least 10 separate structural rearrangements, which include three inversions and at least seven reciprocal translocations. First-generation hybrids between these species are semisterile. Pollen viability is less than 10% and seed viability is less than 1%. *Helianthus anomalus* did not simply combine the chromosomes of its parental species.

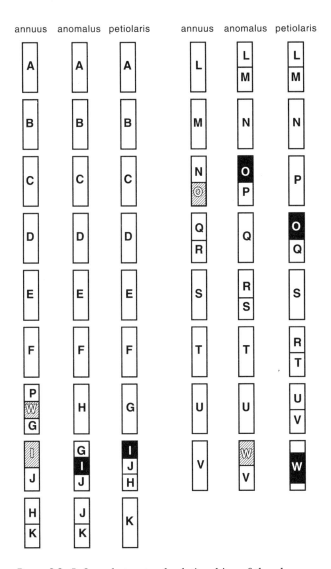

Figure 3.3. Inferred structural relationships of the chromosomes of three *Helianthus* species (after Rieseberg et al., 1995. Redrawn with permission of *Nature*).

Rather it has undergone a minimum of seven novel rearrangements, including a minimum of three chromosomal breakages, three fusions, and one duplication. First-generation hybrids with *H. annuus* have pollen fertilities of 2 to 4% and those with *H. petiolaris* have pollen fertilities of 2 to 58%.

Pleiotropy and Linkage

Mayr (1963) implicitly refers to pleiotropy in the establishment of reproductive isolation in allopatric populations. He states that "the ecological shifts in incipient species are bound to have an effect on their isolating mechanisms. The thesis of the origin of reproductive isolation as a by-product of the total genetic reconstitution of the speciating population is consistent with all known facts."

Given that changes at loci may have manifold effects, there are limitations on their incorporation within populations. Genes of major effect often have large negative pleiotropic consequences and are not likely to be incorporated within a population without corresponding changes at modifier loci (Wright, 1980; Charlesworth, Lande, and Slatkin, 1982; Coyne and Lande, 1985). Thus we should not assume that one allele of major effect will simply (and quickly) replace another.

In view of the prevalence of pleiotropy (Wright, 1977), we would expect many examples in plants. However, this is not the case. Our best understanding of pleiotropy for the same trait across species involves genes governing anthocyanin synthesis. Plants homozygous for an allele conferring white petals in *Phlox drummondii* (Bazzaz, Levin, and Schmierbach, 1982; Levin and Brack, 1995), *Echium plantagineum* (Burdon, Marshall, and Brown, 1983), and in *Digitalis purpurea* (Ernst, 1987) have reduced vigor and competitive ability. The pleiotropic effect of a mutation at an anthocyanin locus is indicated in chimeric plants, where white-flowered sectors show less growth than pigmented sectors (Van Fleet, 1969). The converse also may be true. In *Ipomoea purpurea* white flower homozygotes at the W locus are vegetatively larger and produce more flowers than darkly pigmented homozygotes (Rausher and Fry, 1993). The pleiotropic effect of "white" alleles at anthocyanin loci may be due to the accumulation and diversion of precursors to alternate pathways.

One of the best examples of pleiotropy involves the S locus in *Nicotiana tabacum* (Stebbins, 1959). An allelic substitution that affects petal size and outline also affects leaf, calyx, anther, and capsule dimensions. Another striking example of pleiotropy is in *Arabidopsis thaliana*, where flowering-time genes affect leaf length, number of rosette and cauline leaves, and number of axillary flowering shoots of the main inflorescence (van Tienderen et al., 1996).

In *Lycopersicon esculentum* the gene for suppressed lateral branches also suppresses apical growth and corolla development, and causes deformed anthers and reduction in flower number per inflorescence (Williams, 1960). Alteration of the genetic background through hybridization modifies the pleiotropic effect of the "suppressed lateral" gene. For example, the intensification of the *L. pimpinellifolium* genome through backcrossing brings about a rapid advance toward the normality of the corolla. A shift toward normality in one trait is not necessarily accompanied by such a shift in other traits. This study is important, because the effect of the genetic background on pleiotropy is almost unknown in plants.

Pleiotropy or linkage cause segregation distortion for flower color in *Mimulus*. *Mimulus cardinalis* has red flowers and *M. lewisii* has pink flowers. The difference is controlled by a single gene with pink being dominant to red (Nobs and Hiesey, 1958; Bradshaw et al., 1995). When grown in an environment where few plants die, F_2 progeny of the interspecific cross approximate a $3:1$ ratio of pink- to red-flowered plants. However, when the F_2 were grown at the Stanford garden in the summer, a $1.7:1$ pink: red ratio was obtained (Nobs and Hiesey, 1958). Conversely, when a subset of the same progeny were grown at timberline, an $8.7:1$ ratio was obtained. Nobs and Hiesey also grew a replicate of the F_2s in the Stanford garden in the winter and observed a $4.4:1$ ratio of pink to red.

The distorted segregation ratios were associated with high seedling mortality in environments that are hostile to one or the other species. *Mimulus cardinalis* occurs at low elevations in the California mountains; *M. lewisii* occurs at high elevations under much cooler conditions. High seedling mortality occurs in *M. cardinalis* seedlings when grown at high elevations; such mortality occurs in *M. lewisii* at low elevations (Nobs and Hiesey, 1957). Mortality in F_2 progeny and distorted phenotypic ratios in warm environments ostensibly involves the selective elimination of *M. lewisii* types in warm environments and of *M. cardinalis* types in cool environments.

Evidence for the linkage of genes controlling the contrasting floral traits of related species is forthcoming from *Lycopersicon*. QTLs controlling reproductive behavior and floral traits were mapped in a backcross population of *Lycopersicon esculentum* × *L. hirsutum* (Bernacchi and Tanksley, 1997). The locale on chromosome 1 that contains the S (self-incompatibility) locus also harbors most major QTLs for flower size, flower number, bud type, and inflorescence rachis length. The presence of the H allele at the S locus is associated with more flowers per plant, larger flowers, and longer inflorescences. The S locus also controls interspecific cross-incompatibility. Stigma exsertion is unlinked to the S locus. DeVicente and Tanksley (1993) also mapped a QTL for flower number per plant to the S locus in progeny from a *L. esculentum* × *L. pennellii* hybrid. The association between S locus and floral traits occurs in other families and may be indicative of a common, conserved ancestral gene complex (Bernacchi and Tanksley, 1997).

Another interesting example of linkage has been demonstrated in the recent divergence of the weedy, late-developing *Senecio vulgaris* ssp. *vulgaris* from its nonweedy, early-developing parent, *S. vulgaris* ssp. *vulgaris*. Genetic studies reveal that the major gene for developmental speed is linked to the ray-floret gene, as are chromosomal regions that influence leaf and numbers, and pedicel length (Comes, 1998). This chromosomal region exhibits a substantial proportion of nonadditive gene effects, which contrasts with the traditional additive-polygenic model of evolutionary divergence. A similar pattern occurs within subspecies of *Plantago major*, where the *Pgm-1* locus is associated with many loci affecting ecologically important traits (van Dijk, 1984).

We may now ask how often adaptation and speciation involve major genes. Orr and Coyne (1992) contend that major genes are involved to a greater extent in adaptation and as a basis for interspecific differences than we might judge from theory or qualitative arguments. They reviewed theoretical analyses of macromutational and multifactorial evolution and found that they tend to be biased in favor of the polygenic control of novel adaptations as espoused by Fisher (1958). The usual argument against the importance of major genes in adaptation and speciation is that they are likely to have harmful pleiotropic effects that prevent fixation in natural populations, even though these effects may be overcome by intense selection (Charlesworth et al., 1982; Lande, 1983). Orr and Coyne (1992) conclude that there is little empirical support and inconclusive theoretical support for the notion that adaptation and speciation almost always involve genes of small effect. The data presented here support Orr and Coyne's position.

Genes affecting quantitative trait variation have been mapped in populations derived through interspecific hybridization as discussed above. Such mapping provides the much-needed insights into the genetic basis for species differences that cannot be obtained from classical breeding studies that rely on means and variances from a cross between two populations. Not only will gene numbers be estimated, but it will be possible to discriminate among many genes of small effect and major genes with their modifiers.

Genetic Change and Speciation

There are three pivotal issues related to the genetics of speciation. The first is the amount of change necessary to afford speciation. The second is the nature of differences between ancestor and progenitor. The third is the tempo of genetic change.

Amount of Genetic Change

What can we conclude about the amount of genetic change necessary for the evolution of a new species? This depends on our species concept. If reproductive isolation, or divergence in fertilization systems, alone are the criteria for speciation as in the Biological Species and Recognition Concepts, respectively, then relatively few genes may be sufficient for species formation. If the criterion for speciation is divergence in both reproductive and ecological systems as in the Ecogenetic Species Concept, then genetic change of a greater magnitude is required.

The number of genes responsible for species differences at the time of their divergence almost certainly is much higher than the minimum number sufficient for speciation, because ecological divergence and/or reproductive isolation is not likely to have been achieved by the smallest number of gene changes. Moreover, speciation may have been precipitated by a number of character or genomic changes acting in concert.

Nature of Genetic Differences

Consider next the nature of the genetic differences between ancestor and neo-species. Charlesworth (1992) modeled the fixation of mutant genes in partially self-fertilizing neospecies, and concluded that they would tend to differ from their ancestors in traits that were recessive, because partial selfing facilitates the fixation of advantageous recessive mutants. Dominance relationships of diagnostic traits inferred from crosses between ancestor and partially selfing neospecies support this expectation. If selfing facilitates the fixation of recessive mutants, then we might expect that recessive traits would play a proportionally smaller role in the divergence of outbreeding neospecies.

Another reason recessive alleles will have a smaller role in the divergence of outcrossers is that recessive alleles are less likely to spread in response to natural selection than dominant alleles (Merrell, 1968). In inbreeders there is no theoretical barrier to the spread of recessive genes.

The genetic architecture of character changes has a bearing on the rate of character change. Characters governed by a small number of genes and having a high additive genetic variance tend to spread faster than characters governed by many polygenes and having low heritability and major epistatic effects (Macnair, 1989). Orr and Coyne (1992) also note that a mutation's probability of fixation is directly proportional to its phenotypic effect, and that a major gene contributes more to the character under selection than does a minor gene. This is important because the short-term response to selection depends more on the relative magnitude of gene effects than on the specific number of loci.

The Tempo of Speciation

The time necessary for species divergence is not a simple function of the number of genetic or chromosomal changes involved in the process. Although speciation involving relatively few changes may be rapid, it is not necessarily the case. The population size and strength of selection are also important (Coyne, 1995). Selective change in several traits may be sequential instead of simultaneous (Lande, 1986). If some potentially adaptive genes have negative pleiotropic effects, time may be required for alterations at modifier loci. Stochastically driven change may be very slow (Barton and Charlesworth, 1984), especially if multiple chromosomal changes occur (Walsh, 1982).

The time necessary for the development of postzygotic barriers due to a pair of complementary genes was treated by Orr and Orr (1996) in relation to species subdivision. If alternate alleles were fixed at two loci as a result of genetic drift, the time was not affected by the number of isolated populations that constitute a species. However, if the complementary genes were pleiotropic and affected by natural selection, the reproductive isolation of populations occurs the quickest when a species is split initially into two large populations. Conversely, when postzygotic barriers require stochastically driven change such as chromosomal rearrangements, then reproductive isolation occurs the fastest when population size is small (Lande, 1979a, 1984).

Insights into the speed of postpollination barrier emergence may be gained from studying barriers between selected lines, conspecific cultivars, and cultivars and their wild ancestors. Cultivars pass through genetic bottlenecks and experience periods of inbreeding and directional selection, as do populations undergoing major ecological transitions.

Consider the case of *Phlox drummondii*, which is an endemic annual of central Texas. Seeds were collected by Thomas Drummond in 1835 and subsequently distributed to nurserymen in Europe (Kelly, 1915). Over the next 80 years, over 200 true-breeding lines were established. They differ in corolla size, outline and color, and plant branching pattern and stature. The "Grandiflora" strains were derived from the original material and were the source of the "Dwarf" strains. "Extra Dwarf" strains appeared on the market somewhat later. The most recent strains are 'Twinkles', which are derived from the "Extra Dwarf" strains.

Levin (1976a) measured the level of cross-incompatibility between wild *Phlox* and cultivars. Cross-incompatibility arose as a by-product of cultivar evolution; the greater the number of "phylogenetic steps" between phloxes, the greater is the level of cross-incompatibility. There is a 14% reduction in cross-compatibility when wild *Phlox* is crossed with "Grandiflora" strains, a 20% reduction when it is crossed with "Dwarf" strains, a 35% reduction when it is crossed with "Extra Dwarf" strains, and a 50% reduction when it is crossed with 'Twinkle' strains, relative to cross-compatibility within wild *P. drummondii*. Compatibility was measured in terms of the percentage of pollen grains sending tubes into the pistil.

Whereas rates of barrier building may be obtained for cultivars with known histories and pedigrees, very little information is available for natural populations. One intriguing situation is in *Mimulus guttatus*. Lindsay and Vickery (1967) studied the cross-compatibilities and hybrid fertilities of populations in the Bonneville Basin of Utah using as a natural time clock habitats provided by the recession of glaciers and lakes. The present habitats differ in temperature and length of the growing season. Populations in habitats less than four thousand years old differ among each other in morphology and display substantial reductions in seed-set and reduced hybrid fertility when crossed with other populations. Thus rapid diversification was accompanied by rapid barrier building.

The key question that remains is the tempo of speciation—that is, the time required for both ecological change and the development of at least partial postpollination reproductive barriers. The answer is quite straightforward in the case of allopolyploids, where a shift in ecological profile and barrier building are by-products of hybridization and chromosome doubling. The time is a few generations. This has been demonstrated in the course of hybridization between diploid *Tragopogon* species (*T. dubius, T. porrifolius, T. pratensis*) and the formation of persistent tetraploid populations (*T. miscellus, T. mirus*; Soltis and Soltis, 1993).

The stabilization of diploid hybrid derivatives also may be relatively rapid. Simulation analysis by McCarthy et al. (1995) indicates that recombination speciation is punctuated, with a rapid transition to a stabilized entity following a period of stasis in a hybrid swarm. That period may be hundreds to thousands of generations, depending on the proximity of parental species, breeding system, and the relative fitness of hybrids.

Fertile, stable plant neospecies have been synthesized experimentally in several plant genera, even when F_1 hybrid fertility is low. One experiment involved *Nicotiana langsdorffi* and *N. sanderae* (Smith and Daly, 1959). The F_1 hybrids are semifertile, with pollen fertility near 50%. Three selection lines were derived from the moderately fertile F_2 population, one favoring short corollas like *N. langsdorffi*, another favoring long corollas like *N. sanderae*, and another intermediate corolla tubes. By the 10th generation, the three lines had been stabilized for the selected flower size, several unselected floral and vegetative traits, and normal fertility. The small-flowered line has normal fertility, whereas the other two are partially sterile. The lines have fertility barriers between them. For example, hybrids between the small-flowered derivative and *N. sanderae* have pollen fertilities near 60%, and hybrids between this derivative and *N. langsdorffi* have fertilities between 46 and 58%.

In another experiment Grant (1966a) crossed *Gilia malior* and *G. modocensis*, and obtained a hybrid that was almost completely sterile. Its pollen fertility averaged 2% and its seed fertility averaged 0.007%. Ten generations of inbreeding, and selection for vigor and fertility, yielded a line that was fully fertile with a unique combination of parental traits, and that was intersterile with the parental species. The progression of pollen fertilities across generations is depicted in figure 3.4. It is noteworthy that the vast preponderance of F_2 to F_6 plants were sterile, weak, or both. The chromosome number of the parental species and F_1 hybrid is $2n = 36$. Chromosome numbers in the F_2 generation vary from $2n = 37$ to 40. The stabilized derivative has $2n = 38$. Two other lines selected for vigor and fertility reverted to one parental type.

Ungerer et al. (1998) tested the hypothesis of rapid hybrid speciation in *Helianthus* by estimating the number of generations of recombination to stabilize the *H. anomalus* genome, which was derived from *H. annuus* and *H. petiolaris*. They found that stabilization could occur within 60 generations. The rapid transition may be achieved in spite of the fact that F_1 hybrids are mostly sterile, because in synthetic hybrid lineages pollen fertility recovered to 90% in only four generations of sib mating or backcrossing. This study demonstrates that recombinational speciation in *Helianthus* could have been rapid, not that it indeed was rapid.

Speciation involving chromosomal rearrangements and ecological shifts typically is spoken of as saltational or quantum (Grant, 1981). The speed of the transition remains to be determined. The critical issue here is the time necessary to fix novel chromosomal rearrangements. Chromosomal evolution is dependent on stochastic processes, which typically will not propel a population with the speed associated with selection. Then one requires the confluence of chromosomal change and ecological change. If populations undergoing chromosomal change are small as inferred, the amount of variation in them may be insufficient to support a robust response to selection. Stebbins (1989) suggested that ecological and chromosomal change can be completed during a few hundred or thousands of generations.

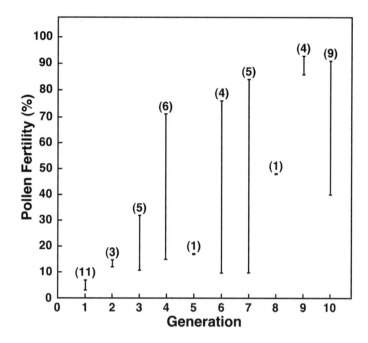

Figure 3.4. The approximate pollen fertility among sister plants in successive generations of inbreeding and selection in *Gilia*. The number of plants tested is in parentheses (based on Grant, 1966).

Speciation associated with an ecological shift and little chromosomal evolution could occur even faster than so-called quantum speciation. Diversification unopposed by serious competitors, such as sometimes occurs on oceanic islands or edaphic islands, may render new species in a few hundred to a few thousand generations. The extensive and rapid radiation of some genera on oceanic islands may not be as surprising as it seems. As mentioned in the previous chapter, major habitat shifts (e.g., the invasion of metalliferous soils) have occurred within the past few hundred years in many plant lineages, as have changes in breeding systems and flowering phenologies.

The Genetic Identities of Species

The degree of divergence in genes controlling reproductive and ecological systems is not necessarily representative of the total genetic changes that have ensued during speciation. Therefore, one would not expect a consistent relationship between taxonomic categories (e.g., species, subspecies) and levels of genetic identity based on loci that were not involved in reproductive isolation.

In some instances the genetic identities of closely related species may be similar to those of conspecific populations (Crawford, 1990).

This situation is illustrated in several genera. Consider the case of *Gaura longiflora* (Onagraceae) and its presumed derivative *G. demareei*. Gottlieb and Pilz (1976) found the same mean genetic identity of 0.99 for pairwise comparisons of populations of the same and different species. Similarly, the mean genetic identities for populations of the closely allied *Heuchera villosa* (Saxifragaceae) and *H. parviflora* are 0.98 and between *H. americanum* and *H. pubescens* 0.99, the same as conspecific populations (Soltis, 1985). Gottlieb et al. (1985) reported mean population identities for the *Layia discoidea* (Asteraceae) and its progenitor, *L. glandulosa*, of 0.90 compared to 0.93 for conspecific populations.

The aforementioned values are based on allozyme loci. High levels of identity have been found between *Mimulus guttatus* and its derivative, *M. micranthus*, for both allozyme and chloroplast DNA restriction fragment length profiles (Fenster and Ritland, 1992). High levels of identity using both approaches also have been found between *Clarkia biloba* and its derivative, *C. lingulata* (Gottlieb, 1974; Sytsma and Gottlieb, 1986).

High genetic identity between species on continents often is not associated with substantial divergence in morphology or ecological profile. However, high identities on oceanic islands often are associated with considerable morphological and ecological divergence. For example, in the Hawaiian composite *Tetramolopium*, the genetic identity among species is 0.95 (Lowry and Crawford, 1985). In Hawaiian *Bidens*, the mean identity is about 0.95 (Helenurm and Ganders, 1985). These values contrast with a mean genetic identity of ca. 0.65 for typical species on continents (Crawford, Stuessy, and Silva, 1987).

The breeding system, through its effect on the genetically effective sizes of species, has an important bearing on the genetic identities of species. In the Hawaiian *Schiedea* (Caryophyllaceae) and its close ally *Alsinidendron*, predominantly self-fertilizing species have lower genetic identities with other species in either genus than do predominantly outcrossing species with similar levels of morphological divergence (Weller, Sakai, and Straub, 1996). Outcrossing allogamous species with small population sizes also tend to have relatively low genetic identities.

Allozyme data often are assumed to be valid descriptors of the overall level of variation within species and genetic divergence between species. This assumption seems valid, given there is typically a good correspondence between allozyme data and other forms of variation, including morphological (Hamrick, 1989). Incongruities between allozyme and other data in neospecies ostensibly signify their rapid evolution through selection and genetic drift. Given that most species traits tend to remain relatively static over time, we would expect allozyme divergence to catch up with the traits by which the species are diagnosed (Crawford, 1990). When genera are considered in their totality, there typically is a good correspondence between allozyme and morphological divergence in the sense that species within different sections of a genus are more divergent in both respects than species within the same section.

Overview

It is evident that major trait differences may be controlled by one to a few major genes or several QTLs, each with minor effects, or a combination of major and minor genes. Of particular interest is the possibility that there may be common genes involved in the same character differences between different pairs of species in the same genus.

Our understanding of the pleiotropic effects of plant genes is woefully inadequate. Theory predicts that genes are apt to have negative pleiotropic effects, and that these effects will retard or oppose a response to selection. The data base is too small even to know whether negative pleiotropic effects are common.

Speciation involves a change in several traits. These transitions may be accomplished by changes in a small number of loci. The number that actually is involved in a given instant remains to be determined. Given that there are a number of character differences between species, we may ask whether the transitions occur synchronously or consecutively. If the differences involve a functional character suite, the transition may be synchronous or nearly so. If the differences involve independent functions, then a temporal relationship is not required.

The Ecogenetic Species Concept calls for ecological divergence and genomic disharmony. As genomic disharmony typically is a by-product of ecological divergence, except when chromosomal changes are involved, it is likely that ecological changes promote the development of genomic disharmony, rather than the other way around. In this respect speciation is both a product and a by-product of adaptive evolution. When chromosome rearrangements are involved as well, then speciation also becomes a product of stochastic evolution, because the fixation of novel underdominant rearrangements typically is based on chance. In this case, genomic disharmony via translocation or inversion heterozygosity may precede an ecological shift in the course of speciation.

We may expect little relationship between the magnitude of character differences between species and the level of differentiation displayed by allozymes or other ostensibly neutral markers. This is supported by the observation that rapid radiation on oceanic islands may yield very divergent species with high levels of genetic identity, whereas moderately divergent species of ancient vintage may have lower levels of genetic identity.

4

The Geographical Scale of Speciation

Speciation must occur within a spatial context. A geographically circumscribed assemblage of individuals diverges from another such assemblage and acquires, through selection and genetic drift, the attributes that render it a new species. That assemblage may range from a single population to a large group of populations. The manner in which species and speciation are studied and how phylogenies and biogeographies are interpreted depends on the geographical unit (or scale) of speciation.

Geographical Models of Speciation

If speciation occurs in an assemblage that is geographically isolated from its progenitor, speciation is said to be allopatric. I recognize three models of allopatric speciation that are loosely based on those described by Brooks and McLennan (1991). Model I is called geographical or vicariant speciation, and involves the physical separation and divergence of two (or more) large groups of populations of the ancestral species. Physical separation is achieved by geological upheaval or major climatic change, on the one hand, or local factors (e.g., the change in the course of a river), on the other. The divergence of the two groups of populations may be accomplished by change in only one of them or by both of them.

Model II, or peripatric speciation, involves the divergence of a population or group thereof just beyond the geographical boundary of the larger ancestral species. Populations beyond the boundary are a result of recent dispersal or are remnants left behind during species contraction. The former typically is the situation envisioned when peripatric speciation is discussed.

Model III, or disjunct speciation, involves the divergence of long-distance disjunct populations from the corpus of the ancestral species. The disjunct populations could be the products of long-distance dispersal or remnants of a once wider species distribution.

Turning to closer spatial relationships, speciation may occur in an assemblage that is in physical contact with the progenitor (Brooks and McLennan, 1991). If contact is on a geographical margin of the progenitor, speciation is said to be parapatric.

If speciation occurs where there is no geographical segregation of populations, it is said to be sympatric. Sympatric speciation may occur in two ways. On the one hand, differentiation begins in spite of gene exchange between the two systems. On the other hand, speciation occurs within a pair of overlapping populations, often as a result of hybridization and polyploidy.

The models of speciation include two components, the spatial relationships of speciating entities and their progenitors, and the sizes (geographical areas) of the speciating entities. Speciation models relate to the aforementioned components in unique ways as outlined in table 4.1. Two types of geographical isolation, proximal and distal, are designated. The former refers to isolation involving relatively short distances, the latter to isolation involving relatively large distances.

Five of the eight possible cells in table 4.1 contain speciation models. Model I (geographical speciation) combines large size with large distances, and so on. No models of speciation combine large size with sympatry or with narrow geographical isolation; and no models combine small size with a contiguous distribution, even though it is possible to have such combinations.

Speciation occurs within a spatial context. If speciation occurs within a small area, involving a population, a metapopulation, or a small group of populations, it is considered to be local. If speciation occurs across a wide-ranging group of populations, it is considered to be regional. Peripatric, disjunct, and sympatric speciation are local.

Whether plant speciation was local or involved the transformation of a large population system has been inferred from the nature of speciational change. If speciation involves chromosomal changes, speciation is thought to be local (Grant, 1981; Levin, 1993). Chromosomal rearrangements and chromosome number changes are promoted by stochastic processes and thus require small population sizes. Polyploidy is initiated locally whether or not it is associated with hybridization. It was assumed that when ecological and associ-

Table 4.1. Speciation modes in relation to the size of the evolving entity and its spatial relations with its progenitor.

		Spatial Relationships			
		Sympatric	Contiguous	Proximal	Distal
Size of Speciating Entity	Small	Sympatric Speciation		Peripatric Speciation	Disjunct Speciation
	Large		Parapatric Speciation		Geographical Speciation

ated phenotypic changes precede the emergence of postpollination barriers, speciation involves the transformation of large groups of populations (Stebbins, 1950; Clausen, 1951; Grant, 1981).

How do we determine whether speciation is a local or regional process? One approach is based on phylogenetic and biogeographic considerations. The different geographical patterns of speciation may have distinctive phylogenetic and biogeographic signatures (Brooks and McLennan, 1991). Geographical speciation yields a tree that is predominantly dichotomous. Peripatric speciation may yield more than one construct. If peripatric speciation were due to random colonization around the margins of the ancestor's range, the tree would contain the ancestor plus polytomies in which the number of terminal species equals the number of peripheral descendants. If colonizers dispersed into new habitats and speciated, and then served as a source of colonizers that repeated the process, the tree would be dichotomous, like the tree from geographical speciation.

The phylogenetic interpretation of the geographical scale of speciation has two major problems. First, we must assume that the progenitor of a new species also changes in its phenotypic and genetic attributes, and thus loses its identity. As will be discussed later, there are many reasons to believe that this is not necessarily the case. Second, interpretations are based on assumptions that may not be valid for a given group of organisms. The history of species may not be adequately reflected in their present genetical and geographical relationships' range. The molecular markers chosen for analysis may not yield trees that accurately reflect species relationships, nor may the most parsimonious trees. Species ranges change in time in an absolute sense and relative to their progenitors' range. Moreover, different spatial patterns of speciation may yield similar phylogenetic trees and biogeographical relationships (Brooks and McLennan, 1991).

With these shortcomings in mind, it is interesting to see some conclusions from phylogenetic-biogeographic considerations. Goldblatt and Manning (1996) inferred that the peripatric and disjunct models of speciation prevail in the African genus *Lapeirousia*. Lowrey (1995) concluded that speciation was largely peripatric in the Hawaiian genus *Tetramolopium*.

The other approach to understanding the geographical pattern of speciation is based on population genetic considerations. I will present arguments in the next section against geographical speciation, and for local speciation.

Geographical Speciation

Ecological and geographically distinct races of plants have been viewed as intermediate stages in the evolution of species for several decades (Stebbins, 1950; Clausen, 1951; Grant, 1981). These races also were viewed as somewhat cohesive breeding communities whose populations evolved collectively. With sufficient divergence, races would become species. The divergence of geographical races is thought to be the primary pattern of plant speciation (Grant, 1981).

According to the theory of geographical speciation, populations within a race diverge collectively from populations in another race. The collective evolution of a race was thought possible because their populations were assumed to be integrated by gene flow (Mayr, 1963). Novel adaptations would spread throughout the range of a race, and the population system would be converted.

The evidence for geographical speciation is largely circumstantial. Races are thought to be transitional stages in the evolution of species because (1) there is no qualitative difference between the types of traits that differentiate races and differentiate species, (2) population systems intermediate between geographical races and species are common in many genera, and (3) partial reproductive isolation may occur between conspecific geographical races (Grant, 1981).

The efficacy of geographical speciation may be sought from a population genetic perspective. Is our understanding of the genetics of speciation compatible with the idea that large geographical races may be transformed into new species? I will argue that the answer is no.

The paradigm of geographical speciation was challenged by Raven (1980, 1986) on the basis that in many, if not most, species gene flow was too restrictive in space to permit races to evolve collectively. In this respect, he reasoned that "no particular pattern in nature logically can be regarded as the precursor of any other; races, subspecies and semispecies cannot be regarded as stages in the evolution of species" (Raven, 1980). Moreover, there are no definite examples of speciation via racial divergence (Stebbins, 1982; Sauer, 1990). Even the fossil record fails to provide such examples (Stebbins, 1987). A manifold change across a large population system is unlikely because there is no effective mechanism to bring it about.

The two mechanisms most often invoked by proponents of geographical speciation are uniform selection across the range of a race and the diffusion of evolutionary novelty by gene flow between populations. There are several reasons why these factors are not likely to be effective.

Uniform Selection

What might happen if habitat differences became more pronounced among races so that all populations within a race would be subject to uniform selection? This in itself is unlikely. However, granting uniform environmental change, let's assume that most populations have the genetic variation to respond. This does not mean that they will respond in the same way.

First, each population represents a semi-isolated sample of the genetic variation of the geographical race with its own unique evolutionary history. Thus populations are likely to differ in the level and character of genetic variation, including epistatic interactions and pleiotropic effects (Cohan and Hoffmann, 1989; Lande, 1983; Endler, 1986; Leroi, Rose, and Lauder, 1994). If populations differed in their covariance matrices, as demonstrated in several plant species (Silander, 1985; Billington, Mortimer, and McNeilly, 1988; Mitchell-Olds, 1986), then divergent responses to selection are even more likely (Lande, 1979b; Via and Lande, 1985). Several genetical processes including natural selection, genetic drift, mutation, and gene flow could contribute to among-population covariance of genetically determined traits (Armbruster and Schwaegerle, 1996). Chromosomal rearrangements also may alter the genetic correlation of traits by creating new linkage groups. Populations also may not respond similarly to uniform selection because they have different patterns of environmental sensitivity or reaction norms, as demonstrated in both greenhouse and field trials in *Phlox drummondii* (Schmidt and Levin, 1985; Schlichting and Levin, 1986) and other species. Given the aforementioned considerations, the response to uniform selection may be the differentiation of populations rather than their shift in the same direction.

The response to uniform selection may be very limited. Genetic correlations between traits related to fitness (by virtue of pleiotropy or linkage) may reduce the response to selection on one trait or trait combinations (Antonovics, 1976). This would happen if alterations in the expression of one character that enhances fitness would be countered by alterations in another that reduced fitness. For example, in *Polygonum arenastrum* there are strong negative genetic correlations between fecundity and age-specific growth. Early commitment to reproduction favors early high fecundity but lower growth rates and fecundity later in life (Geber, 1990). Mitchell-Olds (1996a,b) demonstrated a genetic tradeoff between age and size at first reproduction in *Arabidopsis thaliana* and in *Brassica rapa*. In *Brassica rapa* a line selected for high glucosinolate content was less successful in maintaining fruit and seed production in the face of leaf damage than a line selected for low glucosinolate content (Stowe, 1998).

The response to selection also may be constrained by developmental and functional correlations. Developmental constraints limit the possible developmental states and their morphological expression. Functional constraints limit the values of traits or of trait combinations.

Having noted that linkage of genes, each adaptive in its own right, may retard the response to selection, it is important to recognize that a gene conferring a new phenotype may be linked to a gene(s) that is deleterious. This will retard the response to selection and reduce the probability of fixing a favorable mutation when selection is weak (Barton, 1995). Hitchhiking is less of an impedance on divergence when selection is moderately strong.

I suggest that the same coordinated changes at multiple loci affecting several traits are not likely to occur in different populations. I do not suggest that different populations faced with the same selective pressures cannot respond in kind at the same loci. The independent and thus polyphyletic evolution of a trait involving identical major loci has been demonstrated in *Silene vulgaris*. By crossing heavy-metal-tolerant plants from four metalliferous sites in Germany and one in Ireland with plants from a nonmetalliferous site in The Netherlands, Schat et al. (1996) demonstrated that the tolerance loci for zinc, cadmium, and copper in the Irish plants were identical to those in the German populations. They argue that it is unlikely that dispersal from a single ancestral population could produce the wide geographical distribution of metallophytes in *Silene vulgaris* or other species.

The independent origin of heavy-metal tolerance also has been proposed in the grass *Deschampsia caespitosa*. Bush and Barrett (1993) found that two tolerant populations were clearly differentiated by unique alleles at several allozyme loci. This pattern would be difficult to explain on the basis of stochastic processes or correlated responses to selection following a common origin.

Thus far I have addressed the transformation of a geographical race in terms of a response to uniform selection across its populations. A response to selection is contingent upon genetic variation. Bradshaw (1991) argues that evolution usually is prevented by the lack of appropriate variability. As evidence he cites the common failure of artificial or natural selection to alter the habitat preferences of populations. Populations continually encounter habitats in which they do not occur, but into which their seeds are dispersed, decade after decade. Bradshaw notes that only a *relatively* few species have diversified into several habitats. Neither Bradshaw nor I wish to give the impression that ecotypic differentiation is rare in an absolute sense. There is an abundant literature on this subject (Heslop-Harrison, 1964; Langlet, 1971; Geber and Dawson, 1993).

The presence of strong constraints on alterations in the ecological posture of species (or races for that matter) is seen in their response to long-term environmental change during the Quaternary Period. Such change provides two options for survival—evolving to meet the challenges of a new environment or migrating as a way of staying in the same environment. The analysis of pollen data reveals how tree species have responded to climatic change. From such studies, Huntley (1991) concludes (1) that migration has been the response of most, perhaps all, species faced by Quaternary climatic change; and (2) that the majority of plant species have not responded by adaptive change, at least as far as morphology is concerned. Tree species almost invariably have fossil counterparts that are very similar to them.

Gene Flow

Let us now consider the possibility that novel adaptations originate locally and spread across the remainder of the range. Under this scenario either all novelty arises in the same region or some novelty arises in one region and other novelty arises elsewhere. In either scenario, these adaptations (based on major genes and modifiers or polygenes) would enter a population through pollen or seed dispersal, and soon would be dissociated by recombination. These gene combinations would then have to be reconstituted by selection and increase in frequency before they were transported to the next population, because the probability of a novel gene being randomly transported is a function of its frequency. Different novelties would have different fitness consequences and thus diffuse among populations at different rates. This is an unlikely scenario for racial transformation, especially if there were multiple evolutionary innovations and they originated in different parts of its range.

The rate of novel gene diffusion among populations is a function of gene dispersion and its selective advantage (Fisher, 1937; Slatkin, 1985). The wider the dispersion and the greater the advantage, the faster a gene will spread. Genes that confer a great fitness advantage could spread across a large population system in a few thousand generations, if gene flow between populations was substantial. Conversely, genes with neutral effects or small positive effects would move smaller distances over the same number of generations.

Indirect Estimates of Gene Flow

The average level of gene flow among populations (number of migrants exchanged) may be estimated from measures of gene frequency heterogeneity among populations using Wright's (1969) approximation of the gene flow parameter, Nm, for the island model. Nm is the average number of migrants exchanged between populations per generation. Estimates of Nm for the island model are based on the assumptions that all populations were initially colonized independently from a common source population, that gene flow is random among populations, and that populations have reached a gene flow–drift equilibrium. This model represents the extreme in dispersal over long distances. Nm is estimated using the approximation $Nm = (1 - F_{st})/4F_{st}$, where F_{st} is the standardized variance of gene frequencies among populations. If Nm exceeds 1, gene flow will counteract genetic drift within populations, whereas if it is less than 1, drift will be the prime factor shaping population structure.

When applying Wright's formula to approximate the number of migrants per generation, it must be assumed that (1) a species is at a genetic and demographic equilibrium, (2) in any given generation the rate of gene flow is equal among all populations regardless of spatial location, and (3) the genetic markers are not influenced by selection. These assumptions are unlikely to be met (Bossart and Prowell, 1998). If populations have recently diverged, or extinctions and colonizations have been frequent, or if the species has undergone rapid range expansion, migration and genetic drift may not have reached equilibrium, resulting in under- or overestimates of Nm. Diversifying selection will lead to an underestimate of Nm.

Nm, as estimated from the island model, does not provide a contemporary estimate of gene flow. It is a historic reference point, reflecting the impacts of (1) migration between existing populations, (2) migration affording range expansion, and (3) the replacement of extinct populations on gene frequency heterogeneity among populations (Slatkin, 1987).

With these caveats in mind, it is informative to compare estimates of Nm in species with divergent breeding systems. As reviewed by Hamrick, Godt, and Sherman-Broyles (1995), the mean Nm in predominantly selfing species is 0.57 versus 2.33 for animal-pollinated outcrossing species and 5.73 for outcrossing wind-pollinated species. These values indicate that, in general, populations of selfing species have the potential to diverge through genetic drift, whereas those of outcrossers do not. However, all populations may diverge in response to selective differentials among them.

Gene flow between populations may be restricted by virtue of reproductive barriers. The impact of such barriers on estimates of Nm has been demonstrated in the *Ipomopsis aggregata* complex (Polemoniaceae), where the strength of barriers increases with taxonomic distance (Wolf and Soltis, 1992). Estimates of this statistic range from 1.33 for gene flow between populations within races to 0.017 for gene flow between races within species to 0.010 for gene flow between species.

Gene flow between populations also may be restricted by range discontinuity, as demonstrated in *Pinus*. Species with continuous ranges and their Nm estimates are as follows: *P. ponderosa*, 16.4 (Hamrick, Blanton, and Hamrick, 1989); *P. rigida*, 10.2 (Guries and Ledig, 1982); *P. contorta*, 8.1 (Wheeler and Guries, 1982); *P. banksiana*, 6.7 (Dancik and Yeh, 1983). Pine species whose ranges are strongly discontinuous have much lower values as follows: *P. muricata*, 1.0 (Millar, 1983); *P. halapensis*, 0.6 (Scheller, Conkle, and Griswald, 1985); *P. torreyana*, 0 (Ledig and Conkle, 1983).

What do the aforementioned Nm estimates tell us about species integration? They indicate that populations have drawn genes from others, but they do not indicate that gene flow is sufficient to maintain a species as a unified, cohesive unit.

Direct Estimates of Gene Flow

Estimates of contemporary interpopulation gene flow via pollen may be obtained by analyzing the paternity of seeds (Ellstrand and Marshall, 1985). This analysis is limited to small populations (under 50 plants), unless the population has a very large number of polymorphic loci.

Annual rates of gene flow via pollen into populations of insect-pollinated herbs may be substantial (e.g., 6–18% in *Raphanus sativus*, Ellstrand and Marshall, 1985; Devlin and Ellstrand, 1990; 5–15% in *Lathyrus latifolia*, Godt and Hamrick, 1993; 29–50% in *Asclepias exaltata* Broyles, Schnabel, and Wyatt, 1994). In wind-pollinated trees, gene flow levels may be very high, as in loblolly pine (36%; Friedman and Adams, 1985). Gene flow levels also may be high in insect-pollinated trees, as in the leguminous *Tachigalia versicolor* (25%; Hamrick and Loveless, 1989) and in *Gleditsia triacanthos* (30–50%; Schnabel, 1988). As these values represent gene flow via pollen, they must be divided by two to obtain the rate of gene flow on a diploid basis.

For the purposes of comparison, it is useful to consider a substantial isolation distance treated in several species, in the neighborhood of 1 km. The approximate percentage of seed sired by extraneous pollen varies from 0.4% in *Picea glauca* (Schoen and Stewart, 1987) to 5% in *Cucurbita texana* (Kirkpatrick and Wilson, 1988), 8.6% in *Raphanus sativus* (Ellstrand and Marshall, 1985), and 24% in *Asclepias exaltata* (Broyles et al., 1994).

The values from paternity studies are much higher than values obtained in crops, where the recipient populations contain hundreds of plants (Levin and Kerster, 1974). In crops the target population is genetically distinct from other populations, which makes the identification of immigrants (=hybrids) a simple matter. Isolation distances of less than 1.5 km are sufficient to keep gene flow levels below 0.1% in numerous species (e.g., *Avena barbata, Nicotiana tabacum, Lycopersicon esculentum, Brassica oleracea, Raphanus sativus*, and *Helianthus annuus*).

The difference between the aforementioned native and crop populations is not surprising, given one fundamental difference in their character. The crop populations contained hundreds to thousands of plants, whereas the native populations typically contained fewer than 50 plants. In the case of *Tachigalia versicolor*, the population contained only five individuals (Hamrick and Loveless, 1989). All else being equal, the larger the recipient population the lower will be the rate of immigration. This is so because the proportion of alien pollen in the pollen pool to which a population has access declines as the size of the population increases.

The effect of population size is nicely illustrated in wild radish, *Raphanus sativus*. As noted above, the small wild populations studied have immigrant pollen levels of 6 to 18%, whereas large cultivated populations isolated by about the same distances have much less than 1% immigrant pollen (Ellstrand and Marshall, 1985; Devlin and Ellstrand, 1990). An analysis of gene flow into synthetic radish populations of various sizes also demonstrated that very small populations have higher rates than larger ones (Goodell et al., 1997). In many crops the distances required for varietal purity (less than 0.1% immigrant pollen) are similar to those yielding high levels of immigration in small native populations.

Studies in crop and wild populations of herbs demonstrate two important patterns. First, gene receipt (through pollen immigration) by small populations may be considerable if extraneous pollen sources are nearby. Second, gene receipt by large populations of herbs separated from conspecifics by 1.5 km or more typically is meager.

We know little about gene receipt in small populations isolated by 1 km or more. If we can extrapolate from the mating patterns of widely spaced conspecifics, gene flow could be extensive given that certain pollen vectors are operative. As a case in point consider species of strangler figs (*Ficus*). Pollen dispersal by wasps occurs routinely over distances between 6 and 14 km (Nason, Herre, and Hamrick, 1997). The area from which pollen is drawn usually is too small in plants pollinated by the wind or small insects for pollen dispersal over such distances to occur with any frequency.

Most paternity studies that estimate pollen receipt from extraneous sources have been predominant or obligate outcrossers. If a significant proportion of ovules were fertilized by a plant's own pollen, the impact of pollen from extraneous sources would be much reduced. If almost all ovules were fertilized by a plant's own pollen, the impact of extraneous pollen would be minimal. Moreover, populations of predominately selfing species, by virtue of lower pollen production per flower than outcrossing species, are likely to export much less pollen than their outcrossing relatives.

What do the Nm and paternity estimates say about the diffusion of evolutionary novelty across populations? They indicate that some populations exchange genes at a rate that would readily allow the spread of advantageous genes. The rate of advance of advantageous genes per generation depends not only on gene flow rates between adjacent populations but also (in large measure) on the tail of the gene dispersal curve and the spatial relationships of populations (Slatkin, 1976; Shaw, 1995). The greater the incidence of very long-distance gene flow the faster advantageous genes will advance. Selectively advantageous genes may replace their counterparts in neighboring populations in a few hundred generations (Rouhani and Barton, 1987; Crow et al., 1990), but it might take thousands of generations to replace others throughout a population system (Slatkin, 1976; Livingstone, 1992).

Geographical Isolation

If gene flow between nearby populations is substantial, shall we assume that populations are wedded by contemporary gene flow? The answer depends on the continuity of species' distributions and the distances between populations. If a species were continuously distributed with populations within the pollination and/or seed dispersal range of others, gene flow could be a formidable unifying force (Slatkin, 1985, 1987). If species were distributed discontinuously in local aggregates that were very far from other aggregates, then gene flow could not be an important cohesive force. Nor could gene flow be important if single populations were out of pollination range of other single populations to take the extreme case.

To address whether populations or groups of populations are within pollination and/or seed-dispersal range, we require detailed knowledge about the spatial relationships of populations. One place to obtain such information for many species is the *Atlas of the Flora of the British Isles* (Perring and Walters, 1976). This book contains dot maps based on an aerial grid of quadrats, 100 km^2 in area. Each quadrat is composed of 100 subquadrats 10 km^2 in area. One dot corresponds to the occurrence of a given species in one subquadrat. This dot may represent one plant or several populations. If there are no dots in the adjacent eight 10 km^2 subquadrats, the plants in the target subquadrats are at least 10 km from the nearest conspecifics. If there are no dots in the next ring of subquadrats (16), the plants in question are at least 20 km from the nearest conspecifics.

The two proximity values were obtained for some plant species in Ireland. Many species such as *Plantago major, Dactylis glomerata, Ranunculus repens*, and *Chrysanthemum leucanthemum* are distributed in nearly every quadrat. Thus the mean distance to the nearest quadrat is zero. The distance to the four closest nearest quadrats is close to zero. Immigration might be an operative force in unifying these species.

Other species are common in some areas, but sparse in others. *Glyceria maxima*, for example, occurs in several quadrats in central Ireland, but is infrequent elsewhere. The mean distance to the nearest quadrat with a conspecific is about 3 km. One population or group (=dot) is 80 km from its nearest neighbor. Four were 40 km from the nearest quadrat (figure 4.1). The mean distance to the four nearest quadrats is approximately 22 km. Gene exchange may be an important cohesive factor in some populations, but certainly not in others.

Other species such as *Papaver argemone* are rare throughout Ireland (figure 4.2). The mean distance to the nearest quadrat with a conspecific is about 20 km. Twenty percent of the populations or groups thereof are 40 km or more from the nearest quadrat with a conspecific. If we take the four quadrats nearest to the target quadrat, the mean distance is approximately 30 km. Gene exchange between populations in different quadrats must be exceedingly rare.

Some species occur in juxtaposed quadrats that are separated from other clusters of quadrats by long distance. This is the case in *Spartina townsendii*, which reached Ireland within the past century. The few population clusters are

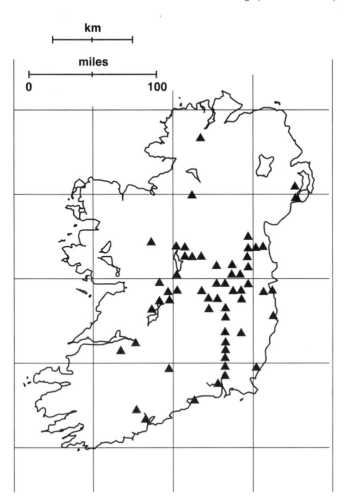

Figure 4.1. The distribution of *Glyceria maxima* in Ireland (from *The Atlas of the Flora of the British Isles*).

about 120 km apart. Immigrants from extraneous clusters must be very rare if not absent.

Given some understanding of spatial patterns, what may be inferred about pollen and/or seed exchange between populations or groups of populations? Pollen receipt from extraneous sources 10 km away or more is likely to be very low in most species. I judge that pollen receipt from populations 20 km away is a rare occurrence in most temperate herbs, and that pollen receipt from populations 30 km away is nil in the vast majority of herbs. Larger distances would be required in most trees to obtain very low migration rates.

We know almost nothing about seed dispersal over distances of 1 km or more. It is probably a rare event in the vast majority of species. Seed exchange

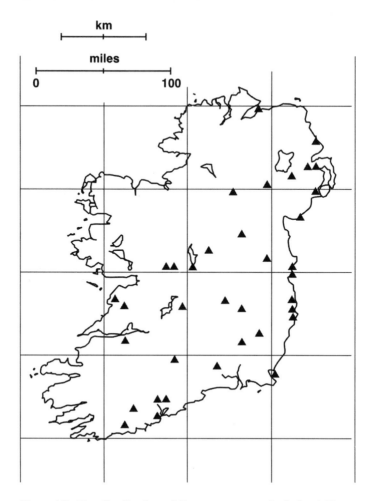

Figure 4.2. The distribution of *Papaver argemone* in Ireland (from *The Atlas of the Flora of the British Isles*).

between populations 10 km apart must be very rare, except in species whose seeds are dispersed by birds, bats, or water. Populations 20 km apart probably are nearly free of seed exchange, except if they employ one of the aforementioned vectors.

A perusal of the *Atlas of the Flora of the British Isles* and other like compilations demonstrates that the potential for interpopulation gene flow varies within and among species. Most species and geographical races therein have substantial range discontinuities that are not readily overcome by pollen or seed dispersal. Therefore, although gene flow may unify some populations, it is unlikely to unify most and certainly not all of them.

Disharmonious Genes and Chromosomes

A developing species often accumulates differences causing genic or chromo-somal disharmony in hybrids with its progenitor. We have to explain how these differences would diffuse across a broad population system by gene flow be-tween populations.

Genes for cross-incompatibility would emerge in some population(s) and have to spread throughout the species in *statu nascendi*. However, crosses be-tween cross-incompatible populations would not be effective. If seeds with the new cross-incompatibility genes were carried to a population, the immigrants would not be able to interbreed with the locals, and the novel genes would be quickly lost.

The diffusion of new chromosomal arrangements across a population sys-tem is even more problematic than that of new genes, because most chromo-somal heterozygotes are partially sterile. New arrangements have a low proba-bility of being fixed in a given population, because stochastic processes play a major role in their fixation (Wright, 1941; Lande, 1979b, 1984). Even if a new translocation were advantageous in a homozygous condition, it is unlikely to diffuse through a population system, because after being introduced into a new population it would be hostage to stochastic processes (Barton, 1979; Lande, 1985).

We have seen that the transformation of widely distributed races into new species is not readily accomplished. Conversely, local speciation does not suffer from any of the problems discussed above. If all of the changes that are associ-ated with speciation occur locally, as argued, then as the species expands its range the resulting populations will share the distinctive phenotypic, genic, cy-toplasmic, and chromosomal features. In essence the expanding new species would be integrated by common descent. This integration would be sustained as the neospecies assumed an ever larger distribution.

As noted earlier, local speciation is thought to be rapid, because strong stochastic processes are involved (Grant, 1981). Whereas this may be so when polyploidy is involved, it need not be so otherwise, especially if speciation in-volves a series of adaptive changes rather than one major stochastically driven change. Indeed local speciation may be very slow (Barton and Charlesworth, 1984), especially if multiple chromosomal changes occur (Walsh, 1982).

Local speciation usually is thought to occur in small populations. Whereas this may be possible, especially when stochastic processes are involved, local speciation need not involve small populations. Adaptive evolution can occur in large populations subjected to new or gradually changing environments (Kirk-patrick, 1982; Milligan, 1986). Even small changes in the environment may be accompanied by the rapid evolution of new tolerances, allocation patterns, and morphological traits (Whitlock, 1997).

Local speciation is not contingent upon change in the ancestral species. The ancestor is likely to retain its character. This scenario is compatible with the theory of punctuated equilibrium, in which the ancestor "suffers" phyletic stasis (Eldredge and Gould, 1972). This scenario also is compatible with the

Ecological Species Concept, wherein stasis is expected to persist as long as the environment remains essentially constant (Van Valen, 1976; Andersson, 1990).

The Unit of Local Speciation

Local speciation in plants (Lewis, 1966; Grant, 1981; Stebbins, 1989) and animals (Templeton, 1981; Mayr, 1982a; Nevo, 1989) is thought to occur primarily within single, isolated populations. Whether speciation in single populations takes hundreds or thousands of generations, these populations must persist long enough for the accumulation of the requisite evolutionary changes. Are single populations likely to persist long enough for speciation to occur in them?

Single Populations

The longevity of single, isolated populations may limit their role as an arena for speciation. As noted by Goodman (1987), "the persistence times of populations subject to substantial environmental variation can be distressingly short." The persistence time depends on demographic and environmental stochasticity and population size. Large populations have a higher probability of surviving 1,000 generations in the face of major environmental and demographic fluctuations than small ones. Small populations, populations subject to large variations in size and populations with a low intrinsic rate of increase (r), are especially vulnerable to extinction.

The longevity of populations is determined in part by birth and death rates, which vary by chance or as a result of random environmental change (Shaffer, 1981, 1987; Lande, 1988). Environmental stochasticity is thought to be the prime cause of extinction in large populations (Goodman, 1987); demographic stochasticity is important only in small populations. The longevity of populations also may be limited by systematic environmental change (biotic or abiotic) that leaves a habitat unsuitable (Harrison, 1991; Thomas, 1994).

Inbreeding depression is another factor that may limit the longevity of populations, especially if they are small, by suppressing population growth rates (Soulé and Wilcox, 1980; Ewens, 1990). Even minor inbreeding depression can elevate the extinction probability above that expected in its absence (Mills and Smouse, 1994).

Inbreeding depression is estimated as $I = 1 - w_s/w_o$, where w_s and w_o are the mean fitnesses of selfed and outcrossed progeny, respectively (Charlesworth and Charlesworth, 1987). Husband and Schemske (1996) report that the level of inbreeding depression across all life stages is significantly higher in predominantly outcrossing species than in predominantly selfing species—0.53 versus 0.23, respectively. The association between mating system and inbreeding depression also is evident among populations within species (e.g., *Clarkia temblor-iensis*, Holtsford and Ellstrand, 1990; *Eichhornia paniculata*, Barrett and Kohn, 1991; *Collinsia heterophylla*, Mayer, Charlesworth, and Meyers, 1996).

A paucity of genetic variation also could have an adverse effect on population longevity by decreasing its growth rate (Allendorf and Leary, 1986; Gilpin and Soulé, 1986). In a population of moderate size, the genetic variance for a quantitative trait can nearly disappear for short periods as a result of genetic drift, and remain at low levels for many generations until replenished by mutation (Burger and Lynch, 1995). Temporary excursions into low levels of genetic variance can threaten the survivorship of populations during periods of environmental change.

The direct or indirect effects of genetic diversity on population survivorship (through the study of offspring fitness) are poorly understood. In *Scabiosa columbaria* and *Salvia pratensis* small populations have less diversity than large ones, but fitness does not vary with size (Bijlsma, Ouborg, and van Treuren, 1994). Conversely in *Gentiana pneumonanthe* fitness is strongly correlated (inversely) with genetic variation and population size (Oostermeijer, van Eijck, and den Nijs, 1994).

The only direct study of the relationship between genetic diversity and population survivorship was conducted by Newman and Pilson (1997) on *Clarkia pulchella*. A series of experimental populations of the same demographic sizes were established, half having a small genetically effective size and half a large genetically effective size. The populations were followed over three generations. Those with low effective sizes have lower germination and survival rates than populations with high effective sizes.

Finally, population longevity may be limited by the accumulation of deleterious mutations, especially if the genetically effective sizes of populations are small. This is because as population size declines, genetic drift becomes a more significant force than selection and the rate of accumulation of deleterious mutations increases (Lynch, 1996). Given that (as in *Drosophila*) mutations are frequent, partially dominant, and individually have small effects, the longevity of populations with effective sizes smaller than 100 typically is on the order of a few tens to a few hundreds of generations (Lynch, Conery, and Burger, 1995). The effective sizes of plant populations often are less, sometimes much less, than 0.3 times the standing crop (Levin, 1977; Heywood, 1986). Thus, plant populations may be particularly vulnerable to extinction via mutational meltdown.

Metapopulations

Given the vulnerability of single populations to extinction, the traditional premise that local speciation occurs within single isolated populations is not readily supported. I propose that the unit of speciation often, perhaps usually, is a metapopulation. A metapopulation is a collection of local populations loosely connected by migration and isolated from the remainder of the species (Levins, 1970). Local populations inhabit somewhat different habitats, and thus exhibit independent demographic traits. Some local populations may be relatively short-lived, whereas others may persist for long periods and be a source of colonists (Pulliam, 1988). When local populations go extinct, they are replaced by immigrants from elsewhere in the metapopulation. The smallest and the most isolated populations have the highest probability of extinction.

The metapopulation model has efficacy for many plant species (e.g., thistles, Olivieri, Couvet, and Gouyon, 1990; thyme, Couvet, Bonnemaison, and Gouyon, 1986; *Pedicularis furbishae*, Menges, 1990; *Eryngium campestre, Medicago falcata*, and *Plantago media*, Ouborg, 1993; *Silene alba*; McCauley, Raveill, and Antonovics, 1995). Short-lived or highly habitat-specific species with good dispersal are the most likely to have well-developed metapopulations (Eriksson, 1996).

Longevity

Like their component populations, metapopulations have finite life spans. The longevity of metapopulations is a function of the counteracting processes of the extinction of local populations and the recolonization of empty patches. The lower the ratio of population extinction to the rate of colonization, the longer the life expectancy of the metapopulation (Carter and Prince, 1981).

In terms of its role in speciation, it is important to note that a metapopulation consisting of a series of populations each of size N will have greater longevity than a single population of size N facing demographic or environmental stochasticity (Harrison and Quinn, 1989; Gilpin, 1990; Hanski, 1991). Even a metapopulation of size N composed of several (x) populations, each of size N/x, may outlive a single population of size N if the environmental variation among populations is not strongly correlated (Hanski, 1989) and if colonization is sufficient (Wissel and Stocker, 1991). A metapopulation also may outlive a local population of similar size because each of the former's components is a source of dissemination, and this allows the metapopulation to better track the environment (Thomas, 1994).

Genetic Structure

The study of the genetic structure of plant metapopulations is at its infancy. However, it is possible to test some expectations from metapopulation models. We would expect younger populations to have greater gene frequency heterogeneity among them (F_{st}) than older populations because the latter would experience gene flow between them. This is indeed what is found in *Silene dioica* (Giles and Goudet, 1997) and in *Silene alba* (McCauley et al., 1995). In the former, populations less than 30 years old had an F_{st} value of 0.057 versus 0.030 for populations between 30 and 280 years old. We also would expect that F_{st} for chloroplast genes would be higher than that for nuclear genes (assuming maternal inheritance) because colonization is effected by seeds. This is indeed what was found in a series of *Silene alba* populations a few generations after the original founding events (McCauley, 1994).

Slatkin (1977) described two modes of population formation that represent the extremes of a continuum. In the "migrant pool" mode founders are drawn at random from all possible source populations. In the "propagule pool" mode founders are drawn from only one source population. The amount of variation in the newly founded population would be less for the propagule pool mode. Which mode best describes what is happening in plant populations? The migrant pool model has no efficacy for higher plants because seed dispersal is restricted (Levin and Kerster, 1974). It is impossible for founders to be drawn from all possible sources. Founders are likely to be drawn from one or a few nearby sources and often are related.

The source of founders is poorly understood. McCauley (1995) estimated the F_{st} value for a group of *Silene alba* populations based on polymorphic allozyme loci and a chloroplast DNA polymorphism. Combining these values with an estimate of the size of the colonizing groups, he could estimate the probability that two alleles found in a colonizing assemblage came from different sources during colonization. He found that most founders came from a single source.

Metapopulations have an organizational structure that is more conducive to the invasion of new habitats than single large populations (Lande, 1980; Wright, 1940, 1982). One avenue for this invasion is Wright's (cf. 1982) shifting balance, which holds that a population may move from one adaptive domain by genetic drift and then be led to an adaptive peak by selection. The novel adaptations would spread through the preferential dispersal of colonists and migrants—that is, interdeme selection. Through this process populations with higher levels of adaptedness would be the largest and have the most emigrants. The smaller the effective size of the metapopulation, the more rapid the transition to a new adaptive peak and the greater the likelihood of its occurrence.

Metapopulation structure may promote selection in some unique ways without invoking Wright's model (Wade, 1996). First, this structure reduces the efficiency of recombination by an amount nearly equal to the amount of subdivision (F_{st}). This in turn makes it easier for selection to favor specific gene combinations. Second, epistatic components of fitness variance, which are not available for selection in large continuous populations, are converted into additive genetic variance within local populations by genetic drift. Thus the potential for local adaptation is much greater in a metapopulation than in a continuous population of the same size. The partitioning of genetic variation among local populations within metapopulations allows interdemic (interpopulation) selection by differential extinction, colonization, and migration to produce large changes in the mean fitness of a metapopulation.

The spatial organization of metapopulations also is conducive to the fixation of underdominant chromosomal rearrangements (Wright, 1940; Lande, 1979a, 1985). With repeated colonization, the rate of chromosomal evolution in a metapopulation is determined by the effective size of local populations. If a metapopulation contains many local populations, several different chromosomal rearrangements may be at various stages of fixation. Thus chromosomal evolution may proceed much faster in a metapopulation than in single large populations. Self-fertilization would further facilitate chromosomal evolution (Charlesworth, 1992).

Phylogenetic Consequences of Local Speciation

Speciation in local isolates is not dependent on phyletic change in the progenitor species. Therefore, the progenitor is likely to retain its character for a considerable period of time. Indeed it is likely to be under strong stabilizing selection for many characteristics (Paterson, 1985, 1986; Eldredge, 1995). During periods of environmental change, the solution seems to be migration to the appropriate environment rather than transformation (Bradshaw, 1991). As noted by Eldredge (1995), "Species represent a level of permanence that acts to conserve adaptive change far beyond the ephemeral capacities of local populations." Like its progenitor, the neospecies also is likely to retain its character, and (like its progenitor) it may serve as the progenitor of another species (Lewis, 1973; Grant, 1981). In contrast, geographical speciation is envisioned to involve differentiation on the part of each isolated population system such that neither descendant species is likely to be identical to its ancestor (Funk and Brooks, 1990).

A locally evolved neospecies belongs to a group that includes an ancestor and some, but not all, of its descendants. Therefore, the ancestor and derivative must be classified as paraphyletic (Donoghue and Cantino, 1988; Neigel and Avise, 1986). Correlatively, many plant species are likely to be paraphyletic (Rieseberg and Brouillet, 1994). This is in contrast to the classical dumbbell model of allopatric speciation that typically generates monophyletic species.

Eventually paraphyletic species become monophyletic (Neigel and Avise, 1986). The transition time depends on the level of geographical differentiation in the mother species, the level of gene exchange between the mother species and the neospecies, and the rates of population expansion and population extinction in the neospecies (Neigel and Avise, 1986). The lower the differentiation and the higher the rates of population expansion, population extinction, and the rate of gene exchange the slower the transition from paraphyly to monophyly. Species with small effective sizes reach reciprocal monophyly faster than species with larger effective sizes. This transition may be of great duration, and thus monophyly may characterize only ancient progenitor species (Rieseberg and Brouillet, 1994).

Neigel and Avise (1986) discussed the phylogenetic status of daughter species with respect to maternally transmitted haplotypes (chloroplast DNA and mitochondrial DNA). Related species arising via the geographical model of speciation were likely to be monophyletic, whereas local speciation may initially yield polyphyletic species that over time become paraphyletic and then monophyletic in matriarchal ancestry. The transition from polyphyly to monophyly would accrue through the stochastic extinctions of cpDNA and mtDNA lineages; and this would take $3n$ to $4n$ generations, where n = the carrying capacity of the larger daughter species (Neigel and Avise, 1986; Pamilo and Nei, 1988). At intermediate times of separation the daughter species would exhibit a paraphyletic relationship. The time to achieve monophyly for nuclear genes would be much greater than for cytoplasmic genes, because the effective size of the former is four times that of the latter (Birky, 1988).

Avise and Ball (1990) extended the work of Neigel and Avise to include the concordance of multiple gene genealogies. Concordances among the genealogical histories of independent loci for two daughter species are likely only when gene exchange between the lineages has ceased for long periods of time. If gene exchange occurs, concordances in gene genealogies as described from a phylogenetic perspective will correspond to that described by reproductive discontinuities. If gene exchange has been severed by geographical barriers, the concordances in gene genealogies may be destroyed unless the lineages were isolated for enough time that reproductive discontinuities evolved.

Paraphyletic species are in conflict with the Phylogenetic Species Concept which holds that species must be monophyletic (Cracraft, 1983). Moreover, the Phylogenetic Species Concept insists that a species cannot remain as the same species after it spawned a daughter species, as it would then become paraphyletic.

Overview

For many years, most botanists and zoologists thought that speciation typically involves the gradual transformation of large populations systems unless there is major chromosomal change. However, increased knowledge about the genetic aspects of speciation, the responses of populations to selection, and the spatial patterns of gene flow among populations leads one to question this paradigm.

If speciation did not involve the gradual transformation of population systems, then conventional wisdom held that speciation occurred within local populations. The problem with this view is that local populations may not persist long enough for speciation to occur. A metapopulation or a cluster of neighboring populations offers a more likely stage for speciation because they persist longer. The greater the opportunity for selection and genetic drift (not concurrent), the greater the likelihood of a substantive genetic change and the origin of a new species.

There is not a prescribed spatial relationship that the speciating entity must have with the parental taxon. The speciating entity may reside near the geographical and ecological boundary of the latter. It may be far removed from the parental taxon, being a remnant of a once wider distribution or the product of long-distance dispersal. It also may reside within the geographical boundary of the parental taxon, but along an ecological margin.

Biogeographical data are sometimes used to infer the geographical scale of speciation and the initial spatial relationship of progenitor and derivative. This practice must be applied with great caution. The spatial relationship between the progenitor and derivative almost certainly will change over a long period of time, and thus the initial relationship may be apparent only if speciation was quite recent. Time also obscures the geographical scale of speciation.

5

The Geographical Expansion of Neospecies

If we track the distributions of successful species, we find that they pass through four major stages. The first stage is one of expansion that ends with the neospecies reaching its geographical limits. This stage is the focal point of this chapter. The second stage is essentially one of macrodistributional stasis, unless novel habitats are invaded. The third stage involves the dissection and/ or movement of a species range in response to major environmental change. The fourth stage involves the contraction and fragmentation of a species range, which when carried to the extreme results in extinction.

Successful neospecies will spread from their sites of origin until they reach impenetrable resistance. This resistance may be in the form of climatic or edaphic conditions, competitors, pathogens and herbivores, or a deficiency of mutualists. The range size achieved within a given time frame depends on when impenetrable resistance is encountered; the earlier the encounter the smaller the range. The factors limiting species ranges are not necessarily permanent. Accordingly, alterations in physical or biotic aspects of the environment may allow neospecies to expand or shift their ranges as conditions permit after a period of stasis.

The Spread of Neospecies

Our first consideration is the spread of neospecies in historical times. I will discuss four species in three genera whose region of origin and spread are well documented.

Tragopogon (Asteraceae) provides some unusual insights into the spread of two tetraploid neospecies. *Tragopogon miscellus* is derived from hybridization and chromosome doubling between *T. pratensis* and *T. dubius*, and *T. mirus* is derived from hybridization and chromosome doubling between *T. dubius* and *T. porrifolius*. The three diploid species were introduced into North America early in this century, and the allotetraploids originated about 60 years ago (Ownbey, 1950). All five species occur in disturbed areas and waste sites within towns. In addition, *T. dubius* and occasionally *T. miscellus* occur along roadsides or in agricultural fields. Ownbey first reported the presence of both tetraploids in the Palouse region of eastern Washington.

Novak, Soltis, and Soltis (1991) reconstructed the expansion of the tetraploids. The range of *T. mirus* increased in the past 40 years from two sites (Pullman and Palouse, Washington) to nine, five of which are in western Idaho. They found about 6,000 plants in Pullman, Washington, and about 2,500 in Palouse that represent more than a hundredfold increase since the populations were first observed in 1949. The spread of *T. mirus* apparently has been promoted by multiple origins at distant sites; long-distance dispersal has played a minor role.

Tragopogon miscellus increased in the past 40 years from 2 small populations (30–35 plants) in Moscow, Idaho, to 38 sites in eastern Washington and western Idaho, many of them having several hundred plants. *Tragopogon dubius* occurs at each of these sites, but *T. pratensis* occurs at only 6 of them. *Tragopogon dubius* is one of the most common weeds in Spokane (Wash.), and is distributed almost continuously several miles into Idaho. *Tragopogon miscellus* also has originated more than once, but long-distance dispersal rather than multiple origins seems to account for its spread. The distributions of both tetraploids are depicted in figure 5.1.

Another composite species of recent vintage is the allohexaploid *Senecio cambrensis*, which is the product of hybridization between *S. vulgaris* and *S. squalidus* (Ashton and Abbott, 1992). The allopolyploid was first discovered in near Wrexham, North Wales, in 1948. Contact between the parental species did not occur until approximately 1910, so the origin of *S. cambrensis* occurred between 1910 and 1948. This species is common in and around the city of its discovery. It also is known from another site in North Wales, ca. 40 km from Wrexham and from Edinburgh, Scotland. Allozyme data indicate that the Welsh and Edinburgh populations have had independent origins.

A third notable species of recent origin is the allotetraploid cordgrass *Spartina anglica* (Gray, Marshall, and Raybould, 1991). The species is the product of hybridization between *S. maritima* and *S. alternifolia*. The first record of the species was in 1892 from the English coastal town of Lymington. The species' spread was rapid, for by the early 20th century it had colonized coastal marshes in many localities in England and had spread to the north coast of France. Because the species is efficient in stabilizing bare mudflats, it subsequently was introduced deliberately and successfully into other parts of Great Britain, Holland, Germany, Denmark, Australia, and China.

Perhaps it is not surprising that all of the examples of neospecies expansion involve polyploids. Only a few "recent" neospecies have been tentatively identified that have evolved independent of chromosomal doubling, and we cannot say with precision when or where they evolved.

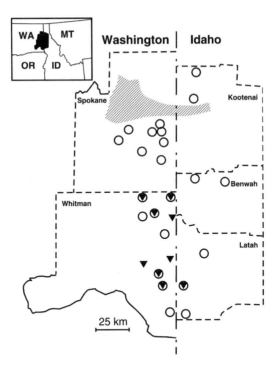

Figure 5.1. The distributions of *Tragopogon miscellus* and *T. mirus* within five counties of Washington and Idaho. The crosshatched area indicates the continuous range of *T. miscellus*. Populations of *T. miscellus* are indicated by circles, and populations of *T. mirus* are indicated by triangles. Both species are present at localities with both symbols (from Novak et al., 1991. Redrawn with permission of the Botanical Society of America).

Principles of Species Expansion

The study of neospecies expansion is limited by the paucity of neospecies whose locale and time of origin is known. However, we may gain insight into neospecies expansion from invading species whose time and site of invasion are known, and from the range expansion of species during the recent postglacial period. Invaders can teach us about expansion from local sites of colonization. Refugial species can teach us about the advance of species over long time periods. Both are germane to the expansion of neospecies across wide spans of hospitable habitats.

The typical pattern for invasions of seed plants and ostensibly for neospecies begins with the slow initial spread in which the species occurs in a few sites. This is followed by a phase of rapid range expansion, and finally a phase in which there is little or no growth (Salisbury, 1953; Mack, 1981; Auld, Hosking, and McFadyen, 1983). The sequence is well illustrated in *Bromus tectorum* (Mack, 1981, 1985). The species was found only at a few scattered locations in western North America between 1882 and 1900. It then expanded rapidly, increasing in area at a rate of 12% per year. By 1930 the expansion had all but ceased. The area occupied by the species is more than 200,000 km^2, and currently it is a dominant species in fields and ranges of the intermountain west. The progression of the species over time is depicted in figure 5.2.

The Lag Phase of Expansion

Not all successful species expand their ranges soon after their origin. Indeed, the aforementioned species of *Tragopogon* and *Spartina* may be anomalies. The historical reconstructions of the invasion dynamics of species from herbarium specimens revealed that there usually is a considerable lag phase between the establishment of local populations and their spread.

Pysek and Prach (1993) documented the spread of *Impatiens glandulifera* (annual), *Heracleum mantegazzianum* (monocarpic perennial, Apiaceae), *Reynoutria japonica*, and *R. sachalinensis* (polycarpic perennials, Polygonaceae) in the Czech Republic (figure 5.3). The four species were introduced at about the same time in the second half of the 19th century, having originally been planted as garden ornamentals. Although these species escaped from cultivation, it was more than 80 years before *H. mantegazzianum* and *R. sachalinensis* began exponential growth. The lag phase took about 40 years in *I. glandulifera* and in *R. japonica*. The number of populations at the end of the lag phase of growth was very low in *I. glandulifera* and *H. mantegazzianum* (12 and 4, respectively), indicating that the exponential phase began shortly after the species had established a few foci in the area. The number of sites at the end of the lag phase was 39 and 22 for *R. japonica* and *R. sachalinensis*, respectively.

Each of the aforementioned species began its exponential growth in the Czech Republic in the mid-twentieth century. The highest rate of expansion was found for *I. glandulifera* followed by *R. japonica, H. mantegazzanum* and *R. sachalinensis*, in that order. The invasion curves of the four species are depicted in figure 5.3. Water was a prime agent of dispersal for all species. The species of *Reynoutria* rarely set seed, so spread by rhizomes must have played a decisive role in their invasion. *Impatiens* and *Heracleum* set abundant seeds, which serve as their dispersal vehicle.

Kowarik (1995) studied the spread of 184 invasive woody species with known dates of first cultivation in Brandenburg, Germany. Some species were first cultivated over 400 years. Six percent of the species began to spread within 50 years, 25% lagged up to 100 years, 51% up to 200 years, and 18% over 200 years. The mean delay in invasion was 131 years for shrubs and 170 years for trees. Several factors might have contributed to the long times (Kowarik, 1995). Foremost were the long juvenile period of woody species, the paucity or inaccessibility of safe sites for germination, and climatic conditions that did not allow the completion of the life cycle. There is no relationship between the lag times and contemporary abundance.

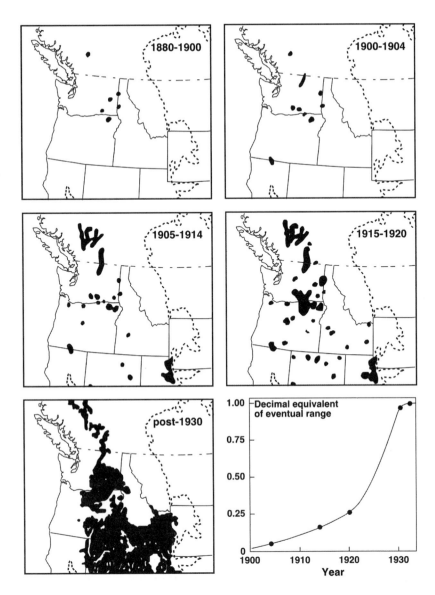

Figure 5.2. Geographical expansion of *Bromus tectorum* in western North America. The graph shows the rate of expansion in proportions of the eventual range (from Mack, 1981. Redrawn with permission of Agricultural Ecosystems and Environment).

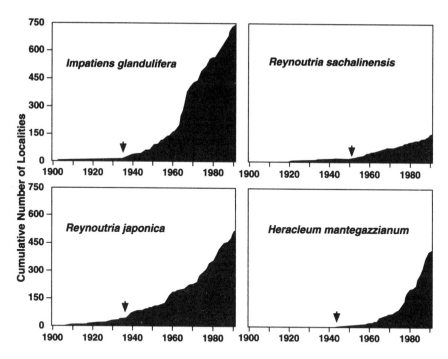

Figure 5.3. Invasion curves of four species in the Czech Republic. The arrows indicate the beginning of the exponential phase of expansion (from Pysek and Prach, 1993. Redrawn with permission of Blackwell Sciences Ltd.).

Rates of Species Expansion

With or without a pronounced lag phase, successful invaders or successful neospecies spread from their regions of introductions or origins, respectively. As discussed by Mack (1985), the rate of species expansion is affected by the (1) number and arrangement of origination points, (2) the amount of habitat adjacent to the origination points suitable for the species, (3) the habitat heterogeneity over longer distances, (4) the availability of corridors and barriers between suitable sites, and (5) the initial size of the original population(s). The rate of expansion also is affected by the (1) population growth rate, (2) length of prereproductive period, (3) local and distant dispersability, and (4) the number and quality of propagules.

Contemporary Species

The rate of species expansion over decades and longer most often has been documented from herbarium specimens. When congeneric species are introduced more or less simultaneously into the same locality, one can compare their rates of spread and area occupied. One can also attempt to identify biological or habitat characteristics that explain a portion of interspecific differences that almost invariably are found.

Weber (1998) documented the spread of *Solidago* species following introduction into Europe from the United States. *Solidago altissima, S. graminifolia,*
and *S. gigantea* were introduced in the 18th century, and were widely distributed by the mid-1800s. The absolute range diameter in *S. altissima* and *S. gigantea* increased logistically from about 1850 to about 1880 with little change thereafter (figure 5.4). Range size is estimated to have expanded 910 km²/yr for *S.
gigantea* and 741 km²/yr for *S. altissima*. The range of *S. graminifolia* increased
little since the beginning of its spread in 1870. Differences among species in
their expansions may reflect differences in dispersal ability. *Solidago gigantea*
has long rhizomes that fragment readily and can be transported by river cur-
rents. *Solidago altissima* has compact rhizomes and occurs in drier habitats. *Solidago graminifolia* appears to rely on seed dispersal for its expansion.

Another study of congeneric expansion involves three *Impatiens* species
that were introduced into the British Isles around 1850 (Perrins, Fitter, and
Williamson, 1993). Single populations of *Impatiens glandulifera* have a maximum rate of spread of ca. 2.6 km/yr compared to 1.6 km/yr and 1.4 km/yr
for *I. parviflora* and *I. capensis*, respectively. For the species as a whole, *I. glandulifera* was estimated to have a maximum rate of spread of 38 km/yr compared
to 24 km/yr and 13 km/yr for *I. parviflora* and *I. capensis*, respectively. The
higher rates of spread of *I. glandulifera* seem due in part to its higher tolerance
to frost, more rapid growth, and higher seed production. All species have ballistic seed dispersal, which only scatters the seeds a few meters.

Forcella, Wood, and Dillon (1986) compared the rates of area occupation
of *Echium plantagineum, E. vulgare,* and *E. italicum* introduced into Australia.
They used herbarium specimens to record presence in 1 degree by 1.5 degree

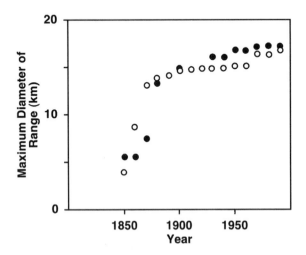

Figure 5.4. The increase in range size of *Solidago gigantea* (open circles) and *S. altissima* (closed circles)
in central Europe (from Weber, 1998. Redrawn
with permission of Blackwell Sciences Ltd.).

grid blocks. In 1910 each species occurred in fewer than 20 blocks. By 1980 *E. plantagineum* was present in nearly 100 blocks, while the other species still occurred in fewer than 20 blocks. The former has become one of Australia's major weeds.

The problem with studying recent range expansion from herbarium specimens is that there is no way to eliminate the human element. This is exemplified by the *Impatiens* invasion. The species have no special mechanisms for long-distance dispersal. The species' ballistic dispersal only carries seed ca. 2 m, and animals only a few multiples of that (Perrins et al., 1993). Humans inadvertently must have assisted their spread by transporting seeds or creating seminatural habitats in which they thrive.

Water may be a very effective vehicle for seed dispersal and thus range expansion in some species. Thebaud and Debussche (1991) documented the range extension of *Fraxinus ornus* after its introduction in 1920 near the Herault River in southern France. The species has wind-dispersed seeds that can float and can be carried long distances by periodic flooding. It spread along the river system at a rate of 970 m/yr. Invasion within a single wetland system by a species with wind-dispersed and floatable seeds also was documented for the woody *Mimosa pigra* (Lonsdale, 1993). This species spread along the Adelaide River (Australia) floodplain at a rate of 76 m/yr. These invasions ostensibly are little influenced by humans.

Holocene Expansions

Another approach to the study of range expansion involves the use of pollen isochrone maps that show the time of first rise in the values of pollen of a specific type, as well as the percentage of that pollen type in a particular sample. Such maps have been used extensively to document patterns of tree spread and tree dominance since the most recent glaciation.

Delcourt and Delcourt (1987) describe the advancement of the northern edge of 20 tree taxa over the past 20,000 years in the eastern and central United States and Canada from isochrone maps. They plot the rate of advance along five equally spaced north-south tracks that run through the following geographical regions: (1) the continental interior along longitude 95 W; (2) the longitudinal axis of the Mississippi Alluvial Valley; (3) the axis from the central Gulf Coast Plain north to Hudson Bay; (4) the northeast-southwest axis of the Appalachian Mountains; and (5) the route along the Atlantic Coastal Plain into the maritime provinces of eastern Canada. These data are summarized in table 5.1. Rates of advance vary from 45 to 287 m/yr.

Given that the mechanism of dispersal is known for each taxon, there might be a simple relationship between dispersal vector and migration rate. Wind and animal dispersed seeds are expected to be carried the farthest (Willson, 1993). Although the three taxa with greatest mean migration rates (*Betula, Salix,* and *Populus*) are dispersed primarily by the wind, there is no simple relationship between dispersal mechanism and migration rate (Delcourt and Delcourt, 1987).

Climatic tolerance and the availability of suitable habitats ostensibly override any inherent intertaxon differences in dispersability. The importance of

Table 5.1. Migration rates for the leading (northern) edge of distribution (from Delcourt and Delcourt, 1987).

Taxon	Overall Rate Averaged for 5 Tracks (m/yr)	Minimum Overall Track Rate in m/yr (Trk#)
Salix	287	181 (1)
Populus	263	167 (1)
Betula	212	195 (4)
Tilia	209	116 (3)
Tsuga	202	113 (3)
Larix	189	119 (2)
Fagus	169	95 (1)
Abies	159	110 (1)
Picea	141	104 (3)
Juglans	140	70 (4)
Ulmus	134	123 (1&2)
Acer	126	80 (2)
Quercus	126	110 (2)
Fraxinus	123	94 (1)
Carya	119	74 (5)
Nyssa	70	42 (2)

these factors is reflected in the differences in migration rate of a taxon from one track to another. In general, the rate of migration was fastest along the easternmost track. Their importance also is reflected in the temporal variation in migration rate of each taxon. The highest migration rates for most taxa were between 8,000 and 16,000 years ago.

Estimates of the rate of spread of trees over the past 10,000 years have been made for the European mainland (Huntley and Birks, 1983) and for the British Isles (Birks, 1989). These data are summarized in table 5.2. Several taxa have an average rate of migration exceeding 300 m/yr. Rates of spread also vary in time, being two to three times higher in some taxa 7,000 to 10,000

Table 5.2. Estimated migration rates (m/yr) of representative trees in Europe and North America.

Taxon	U.K.[a]	Europe[b]	North America[c]
Betula	250	more than 2000	195–232
Ulmus	550	500–1000	123–162
Quercus	350–500	150–500	110–163
Pinus	100–700	1500	105–170
Fagus	100–200	200–300	95–214
Fraxinus	50–200	200–500	94–177
Tilia	450–500	300–500	116–357

[a]from Birks, 1989
[b]from Huntley and Birks, 1983
[c]from Delcourt and Delcourt, 1987

years ago than in recent millennia. Rates of spread apparently decline as trees approach their range limits.

The spread of trees across terrain liberated by the retreat of the glaciers may be studied in a more recent context. The advancing front of tree species has been studied in Glacier Bay, Alaska, following the rapid retreat of glaciers there between 1750 and 1840. Fastie (1995) found that several species migrated across 50 km of deglaciated terrain at rates of 300–400 m/yr, which are similar to early Holocene migration rates in western North America.

Although much is known about trees, it is clear that North American and European herbs also have undergone a rapid expansion northward during the Holocene. Numerous species moved from 450 to 2000 km during the past 16,000 years. Their rates of migration may have been as rapid as those by wind- and animal-dispersed trees. This assemblage includes species such as *Asarum canadense*, whose seeds are ant-dispersed and have a mean dispersal distance of 1.54 m (Cain, Damman, and Muir, 1998). Unfortunately, herb migration rates cannot be calibrated per species from the fossil pollen record.

We have considered migration rates of expanding species from a contemporary perspective and from a historical perspective. Rates of expansion from areas of introductions or glacial refugia may be relatively rapid when species are crossing favorable terrain. The differences between the introduced and refugial species are surprisingly small, given that the latter are all trees. This suggests that both groups of species offer valuable insights into the expansion of neospecies across a hospitable landscape.

The mean migration rates of introduced and refugial species often greatly exceed their mean seed dispersal distances, especially when we take into account the fact that many years are required for perennials, especially trees, to reach reproductive maturity. With the exception of trees with wind-borne seeds and herbs with dust-like seeds, most seeds are dispersed within a few meters of their source (Willson, 1993; Cain et al., 1998). Long-distance dispersal is the key to reconciling the disparity between mean dispersal distances and rates of migration.

The Mechanics of Species Expansion

Most range expansions do not proceed along one radiating or advancing wave front. Rather, invaders usually radiate from several foci, originating either from multiple introductions or long-distance dispersal early after single introductions (Bannister, 1965; Baker, 1986). The rate of expansion is strongly affected by the number of foci—that is, the number of populations exporting seeds. Each additional population accelerates this rate, even when only the number of foci and not the total area initially occupied is increased (Mack, 1985; Moody and Mack, 1988). Each new population increases the size of the expanding margin of the invasion more than the mere expansion of the original population. Expansion also is enhanced by the establishment of foci that are far apart (Auld et al., 1983). Eventually the local radiating clusters of populations coalesce.

Multiple Foci

Baker (1986) described radiation from multiple foci in several California species. Early stages of invasion are exemplified by *Chondrilla juncea* (Asteraceae), which has a focus in north-central California and scattered satellite populations in coastal, central, and southern California. Similarly, *Helianthus ciliaris* has one main focus (in the Los Angeles area), but also has satellite populations at several localities throughout California. The coalescence of populations radiating from multiple foci is seen in *Centaurea repens*.

The rapid spread of trees through unglaciated areas over the past 15,000 years probably was facilitated by the continuous establishment of multiple, well-spaced foci. The process may be visualized in hemlocks at their northwestern border in Wisconsin and adjacent states. The fossil record shows that continuous populations of hemlock reached their present limit in Wisconsin 1,800 years ago, followed by the establishment of a large outlier 20 km west of the limit 1,300 years ago (Davis, 1987). Other scattered outliers have been detected from pollen sediments in Michigan. Today there are several outlying populations in Minnesota of recent vintage, the farthest being 150 km beyond the species limit. Other outliers of unknown age occur in Indiana and Ohio.

The importance of foci number has implications for the early spread of neospecies. Neospecies that evolve in a metapopulation system or in a system of neighboring populations, and thereby have multiple sites of dispersal, may expand faster than neospecies that have a single population or location, even if the population is large, because most seeds will be dispersed within or near the population rather than to new sites. Neospecies with multiple origins, and thereby originating at different times and different places, are likely to expand faster than a neospecies arising once. The rapid spread of the neospecies *Tragopogon miscellus*, as described above, surely was fostered by its multiple origins. It may have originated as many as 21 times (Soltis et al., 1995).

Long-Distance Dispersal

Analytical models have been formulated to understand and predict range expansion. The first models, now referred to as neighborhood diffusion models, treated population expansions as an advancing wave (Skellam, 1951; van den Bosch, Zadocks, and Metz, 1988). These models followed Fisher's (1937) influential model that portrayed the spread of an advantageous gene as an advancing wave. In such models dispersal is characterized by the variance in parent-offspring distance. These neighborhood diffusion models often are inadequate descriptors of the spread of invading species because they assume normally distributed dispersal distances. This assumption fails to take into account the rare but important dispersal events that carry seeds far beyond the boundary of populations.

Propagules in the tail of the distribution are responsible for colonization events. Portnoy and Willson (1993) studied the shape of seed-distribution tails in many species with a range of dispersal devices. They found that many species' distributions had algebraic tails. They predicted that wind-borne, vertebrate-dispersed, and ballistic seeds are the most likely to have algebraic tails. A tail with an algebraic form has a greater reach than a tail with an exponential form. Thus for a given distribution of "safe sites" not markedly aggregated, seed shadows with algebraic tails have the potential to reach more safe sites than those with exponential tails.

Stratified diffusion models, which take into account short-distance and long-distance dispersal, now are being formulated to describe and explain the speed and pattern of range expansion. Migration by short-distance dispersal expands the area inhabited from the periphery, while long-distance dispersal generates new populations far from the resident range. Shigesada, Kawasaki, and Takeda (1995) and Kot, Lewis, and van den Driessche (1996) have shown that the rate of range expansion is critically dependent on the distance of short- and long-range colonization and the incidence of long-distance colonization. Rare long-distance events can greatly accelerate the rate of species expansion. These events have the greatest impact on species with low population growth rates. Their results explain various types of nonlinear range expansions observed in biological invasions. The character of the tail may be as important as the mode of a distribution in explaining rates of postglacial advance in many plant species (Portnoy and Willson, 1993).

Both stratified and neighborhood diffusion models predict that range expansion will be radial because they assume that the environment is homogeneous. In a heterogeneous environment the rate of spread depends on attributes of the environment. As discussed below, rates of spread are shown to be quite sensitive to regional conditions.

Given that we know the shape of the dispersal curve (with the exception of the tail), and the mean distance a species has migrated over a long time frame, it is possible to estimate the incidence of long-distance dispersal. Consider *Asarum canadense* (Aristolochiaceae) that has traveled over 200 km in 16,000 years. Cain et al. (1998) constructed models that examined the tail of the seed dispersal curve. They concluded that more than 1 seed per thousand would have had to disperse more than 1 km to explain the northward Holocene expansion of this species in the United States. The mean observed dispersal distance is 1.54 m.

Another attempt to estimate the incidence of long-distance dispersal was made by Clark et al. (1998) on several tree species. His models suggest that 2 to 10% of seeds are dispersed long distances, which he defines as 1 to 10 km. This is for animal- and wind-dispersed seeds.

The models of Clark et al. and Cain et al. depend on population growth parameters as well as those of dispersal, and thus provide only a very rough estimate of the incidence of long-distance dispersal. Nevertheless, their estimates are surprisingly high.

The incidence of long-distance dispersal depends on the number of seeds or fruits that are exported from a population and the properties of those propagules. The probability that at least one propagule reaches a specified distance from the source is proportional to the number of propagules dispersed (van der Plank, 1960). The specific number ostensibly will be a function of the population size and the seed dispersal curve, especially the tail of the distribution.

Species vary in their adaptations for dispersal and the seed shadows they generate. Long-distance dispersal apparently is most common in species with succulent fruits that are eaten by animals or seeds that adhere to their exteriors, and in species with very light seeds that are wind-dispersed (Willson, 1993). Jays and other birds are thought to be the agents of long-distance dispersal in oaks and other large-seeded plants (Johnson and Webb, 1989; Clark et al., 1998). Darley-Hill and Johnson (1981), and Mattes (1982, cited in Poschlod and Bonn, 1998) reported that 54% and up to 60%, respectively, of the entire seed production of oaks and *Pinus cembra* can be disseminated by different species of jays. Large herbivores also may be important in the long-distance dispersal of seeds (Janzen, 1981; Malo and Suarez, 1995) as may carnivores (Herrera, 1989).

Whereas endozoochoric dispersal may be the norm, ectozoochoric dispersal also may be important in long-distance transport, as recently demonstrated by Kiviniemi (1996). Even species whose seeds are adapted for wind dispersal may be transported long distances by birds (Wilkinson, 1997). The maximum migration rates during the Holocene do not differ significantly between animal- and wind-dispersed trees (Delcourt and Delcourt, 1987).

Our understanding of the rate and process of species expansion may be applied to the expansion of neospecies from their centers of origin or to their spread once they have achieved a substantial range. If we knew enough about the dispersal of neospecies, their habitat preferences and availability, and their biotic interactions, it would be possible to predict their future course of migration. It also might be possible to show that some neospecies had reached their limits.

Ecological Attributes of Expanding Species

There are so few neospecies that have been followed in historical times that it is impossible to find any common biological attributes that explain their success. However, expanding neospecies share a common feature: they move into areas they have never occupied—that is, they are invaders. Can we learn something about the potential for neospecies invasions from the literature on invasive plants? The answer definitely is yes.

Considerable attention has been given to attributes of invaders collectively (Baker, 1974; Pysek, Prach, and Smilaver, 1995) or within single genera (e.g., *Eupatorium*, Baker, 1965; *Pinus*, Roy, 1990; Rejmánek and Richardson, 1996). Rapid population growth, high reproductive potential, adaptations for short- and long-distance dispersal, unspecialized pollinators or self-compatibility, short juvenile period, phenotypic plasticity, and broad ecological tolerance are mentioned often among attributes of invaders. In spite of the multitude of studies on introduced species, predictability of which species are likely to become successful invaders is poor (Crawley, 1987; Roy, 1990). The complexity of the interaction between the species and the community contributes substantially to this state of affairs. So where does this leave us with regard to neospecies? The attributes of neospecies are likely to have little predictive value regarding their success and rate of spread in a given region.

If we consider that all expanding species are invaders, then we would not expect to formulate a list of features that characterize invaders. Expanding species embrace a multitude of trait combinations, even those that might be considered ill-suited for invaders. However, all expanding species have properties that allow them to invade some community(ies). All communities are invasible (Crawley, 1987). What is required is the effective exploitation of local resources. A key resource may be seed dispersers. For example, the success of woody species invading primary tropical forests is increased if they have fleshy fruits that are dispersed by vertebrates (Rejmánek, 1996).

Genetical Aspects of Species Expansion

The genetic attributes of invasive species have been studied with the hope of finding strong correlations between variables such as ploidal level, genetic diversity, heterozygosity, and mating system on the one hand and invasibility on the other. There seem to be none (Barrett and Richardson, 1986; Gray, 1986; Brown and Burdon, 1987). Some invading species are diploid, others are polyploid. Some invaders have substantial genetic variation, others have little. Some invaders are predominantly self-fertilizing, others cross-fertilizing, and others predominantly asexual.

The Imprint of Bottlenecks

The process of invasion per se may have a characteristic genetic signature. Genetic patterns may be produced that can persist for hundreds or thousands of generations. If the numbers of founders of the outlying populations are small, these populations will undergo bottlenecks, and therefore may have reduced genetic diversity relative to interior populations. Correlatively, gene frequency heterogeneity among populations may be greater near the edge than in the interior of the species range (Hewitt, 1989, 1993; Nichols and Hewitt, 1994).

The imprint of founder effects is best sought in the cytoplasmic genome, because the effective number of organelle genes is one-fourth that of nuclear genes when the sex-ratio of breeding individuals is 1:1 (Birky et al., 1983). If gene flow by seeds is more restricted than that of pollen, and if cytoplasmic genomes are maternally inherited, cytoplasmic genomes also will be more sensitive to stochastic processes subsequent to colonization (Ennos, 1994; McCauley, 1995).

Population Systems

Computer simulations of an advancing front generated by short-distance and occasional long-distance dispersal show that founding events will generate a high level of genetic differentiation among populations for maternally inherited genes, which is likely to persist for long periods of time (LeCorre et al., 1997). Populations behind the front will expand and spawn new populations in the neighboring area. This will yield genetically distinct clusters of populations that eventually will exchange genes with each other. The process of long-distance colonization and diffusion behind the front will produce long-lasting gene frequency clines following the direction of species expansion (Sokal and Menozzi, 1982; LeCorre et al., 1997).

Hewitt (1989, 1993) and Nichols and Hewitt (1994) argue that if the numbers of founders of the outlying populations are small, these populations will undergo bottlenecks, and therefore may have reduced genetic diversity relative to interior populations. Correlatively, a series of outlying populations may by chance lack rare genetic variants and be fixed for the one most common behind an advancing front. When these populations generate new outlying populations and other populations between them, the advancing front of a species will have no variation for the loci in question.

Nichols and Hewitt (1994) also argue that rapid expansion can yield patterns that persist for hundreds or thousands of generations. The same geographical pattern initially will be displayed by both chloroplast and nuclear genomes. However over time the nuclear patterns will become much less distinct, especially in species in which pollen dispersal is much more wide-ranging than seed dispersal.

The amount and organization of the variation to be partitioned depends in part on the number of refugia from which populations are spreading and the distance between them (Hewitt, 1996). The more refugia and the closer they are, the more variation to be partitioned and the finer the scale of haplotype dominance. Plant species often radiate from at least two major refugia, the location of which varies among species (Taberlet, Fumagalli, and Wust-Saucy, 1998).

The expectations of Hewitt and Nichols have been realized in several species. Petit et al. (1997) studied the spatial distribution of chloroplast DNA polymorphisms in the closely related *Quercus petraea* and *Q. robur*. The study area covered approximately 60,000 km^2 in western France. The same haplotypes were shared by the two oak species in 101 of the 125 forests where both species were sampled. Autocorrelation analyses show the presence of strong genetic structure for the three most frequent haplotypes, which was significant up to 30 to 40 km and was very similar for both species (figure 5.5). There are patches of several hundred square meters where only one haplotype occurred.

The large size of the chloroplast patches indicates that rare long-distance dispersal events probably were involved at the time of colonization, about 10,000 years ago. The localized systematic sharing of haplotypes by the two oaks strongly suggests that they were exchanging genes as they spread northward and westward across Europe.

Long-distance dispersal also appears to have shaped patterns of genetic variation in European *Fagus sylvatica*. Demesure, Comps, and Petit (1996) sampled 85 populations in an area from western France to the Ukraine and Crimea, and from Sweden to Sicily. They identified 11 cpDNA haplotypes. Only the most common haplotype occurred in northern Europe. Most of the variation was found in Italy, Crimea, and in the Pyrenees. This pattern is consistent with previous results found for allozymes. Gene frequencies at several loci were more asymmetrical in northern Europe than in southern Europe. Fossil pollen data indicate that there were two main refugia in southern Europe from which beech spread to the north and west: Italy and the Carpathes (Huntley and

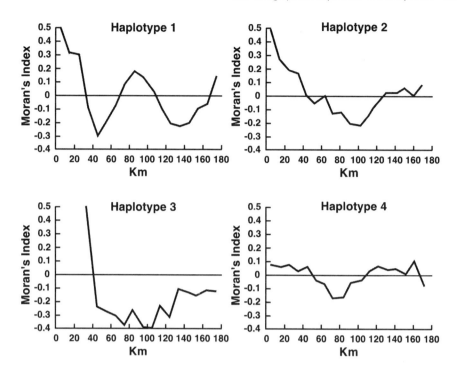

Figure 5.5. The spatial genetic structure of the four most frequent cpDNA haplotypes in *Quercus petraea* (from Petit et al., 1997. Redrawn with permission of The National Academy of Sciences, USA).

Birks, 1983). The latter appears to have been the primary source of the postglacial beech expansion.

The cpDNA and allozyme studies on beech also provide insights into the relative importance of pollen and seed flow in countering the dispersive effects of sampling error and other processes on the gene frequencies of populations. The low level of differentiation for nuclear genes ($G_{st} = 0.05$; Comps et al., 1990) contrasts with a high level of differentiation for cpDNA ($G_{st} = 0.83$; Demesure et al., 1996). Given maternal inheritance of cpDNA in beech, pollen flow between populations must occur at a much higher rate than seed flow.

Alnus glutinosa shows a somewhat similar pattern of cpDNA haplotype diversity. King and Ferris (1998) identified 13 haplotypes that showed a high degree of structuring on a European scale. The pattern indicated that most of central and northern Europe was colonized from one refuge in the Carpathian Mountains. The level of population differentiation for nuclear markers ($G_{st} = 0.24$; Prat, Leger, and Bojovic, 1992) was much less than that for chloroplast markers ($G_{st} = 0.87$) indicating that gene flow via pollen was much higher than gene flow via seeds.

Differences in the level of interpopulation gene flow via pollen and seeds may be estimated from gene frequency heterogeneity among populations (F_{st}

or G_{st}), if data are available for both nuclear and maternally inherited markers (Ennos, 1994). Values in the literature are expressed as the ratio of pollen to seed gene flow. The values vary from 1.8 in *Eucalyptus nitens* to 500 in *Quercus petraea* (table 5.3). The values are 84 for the aforementioned *Fagus sylvatica* and 23 for *Alnus glutinosa*. The greater the ratio of pollen- to seed-mediated gene flow, the greater the expected differential in the geographical structuring of nuclear and chloroplast markers.

The tree species mentioned above show a loss of genetic variation during their northward migration in the Holocene. Other tree species also display this pattern, such as fir *Pseudotsuga menziesii* (Li and Adams, 1989) and *Pinus contorta* (Cwynar and MacDonald, 1987).

The loss of variation during migration away from refugia also has been demonstrated in herbs. Broyles (1998) reported that northern populations of *Asclepias exaltata*—that is, those occupying previously glaciated regions—possess significantly fewer polymorphic loci, alleles per polymorphic locus, and expected heterozygosity than populations in the southern Appalachian refugium. Nineteen rare alleles observed in this refugium were absent in the northern populations.

Migration from refugia also has been accompanied by a decline in genetic variation in *Silene regia*. Dolan (1994) found that allozyme variation declines significantly along a transect running from populations in Arkansas and Missouri to glaciated regions of Ohio.

Lower levels of variation near an advancing edge are not invariably a simple consequence of migration dynamics. Populations in refugial areas today may be more variable because of introgression between divergent refugial populations. This process may explain higher levels of variation in southern U.S. populations of *Liriodendron tulipifera* than in northern populations (Parks et al., 1994).

Populations in refugial areas don't invariably have higher levels of genetic variation, especially if they have contracted and have been isolated for long

Table 5.3. The relative rate of pollen flow to seed flow.

Species	Pollen/seed ratio	Reference
Eucalyptus nitens	1.8	a
Argania spinosa	2.5	a
Hordeum spontaneum	4.0	b
Pseudotsuga menziesii	7	a
Alnus glutinosa	23	c
Pinus contorta	28	a, b
Pinus radiata	31	b
Pinus attenuata	44	b
Fagus sylvatica	84	a
Quercus robur	286	a
Quercus petraea	500	a

[a]El Mousadik and Petit, 1996
[b]Ennos, 1994
[c]King and Ferris, 1998

periods. This is illustrated in European populations of *Saxifraga cernua*. Individual populations in the Alps have no variation for a series of RAPD markers, and populations within the same geographical cluster share the same unique phenotype (Bauert et al., 1998). This contrasts with the substantial intrapopulation variation in Scandinavian populations.

As a species advances, populations that were once on the leading edge may find themselves hundreds of miles from it, and in turn may be subject to different selective pressures than when they were at the front. This is seen in *Pinus contorta*, where the times since the foundings of populations have been estimated by radiocarbon dating. Cwynar and MacDonald (1987) showed that both seed mass and wing loading are positively correlated with population age (figure 5.6), which means that populations founded more recently have seeds that are more dispersible.

Given that long-distance colonization from the leading edge of an expansion sets the stage for stochastic processes to impact the attributes of the advancing species, the possibility exists for the fixation of novel chromosome arrangements (Hewitt, 1993). A population with a novel arrangement may then export migrants into locations not previously inhabited by the species, thereby giving rise to a chromosomally distinct population system. This is apparently what happened in *Clarkia rhomboidea* (Mosquin, 1964). This species is an allotetraploid with two common chromosomal arrangements (Northern and Southern). The arrangements differ by two translocations involving three pairs of chromosomes. Populations also differ in paracentric inversions. The Northern arrangement is thought to be the original one because it occurs in north-central California, where both putative parents occur. The arrangement occurs northward to Washington and westward into Idaho and Utah. The Southern arrangement is found in the southern Sierra Nevada, southern California, and Arizona. Other arrangements also have been found (figure 5.7).

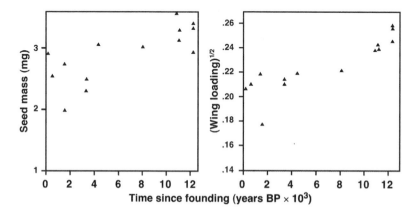

Figure 5.6. Seed mass and square root of wing loading of *Pinus contorta* populations in relation to the time since founding (from Cwynar and MacDonald, 1987. Redrawn with permission of the University of Chicago).

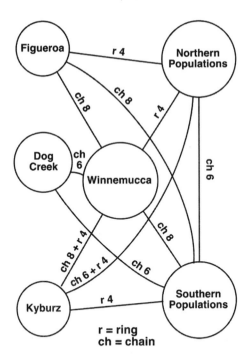

Figure 5.7. Summary of maximum meiotic prophase configurations in F_1 hybrids between populations of *C. rhomboidea* (from Mosquin, 1964. Redrawn with permission of *Evolution*).

Disjunct Populations

If stochastic processes are important in shaping variation, it should be evident in disjunct populations. Unfortunately, almost nothing is known about cytoplasmic genetic diversity in disjunct populations relative to populations within the body of the species. However, genetic diversity contrasts have been made in several species using allozymes. They show that disjunct populations typically have less variation and are more differentiated *inter se* than interior populations (e.g., *Erythronium montanum*, Allen et al., 1995; *Lilium parryi*, Linhart and Premoli, 1994; *Eichhornia paniculata*, Glover and Barnett, 1987).

The effect of stochastic processes is evident in four isolated pioneer stands and the related source population of *Nothofagus menziesii* in the South Island, New Zealand (Haase, 1993). The age of the source population probably is more than 10,000 years. Two of the pioneer populations are in the neighborhood of 2,000 years old, and two populations are less than 600 years old. The youngest pair of populations has the lowest levels of allozyme variation relative to the source population and are the most divergent from it.

The youngest and smallest of the *Nothofagus* populations (Otira Valley) have the greatest genetic affinity with one of the 2,000-year-old populations

(Spray Creek), from which it presumably was derived following long-distance dispersal. Based on the allele frequencies at Otira Valley and on plant size, Haase (1993) inferred the genotype (at 5 loci) of the single founder and the genotypes of 16 first generation and 44 second generation plants. During the three "successive generations" the frequencies of 12 of the 16 alleles investigated shifted from those of the pioneer tree to those of the presumed source population. This was attributed to long-distance pollen flow.

The relationship between population age and their genetic attributes (based on isozyme surveys) has been studied by Giles and Goudet (1997) in *Silene dioica*. The populations were on the islands of the Skeppsvik Archipelago in Sweden. Population age was judged by the emergence times of islands from the Baltic Sea and the successional status of the species. Newly founded populations are more divergent *inter se* than older populations, and newly founded populations have less genetic diversity than older ones. There is no correlation between genetic and geographical distances for young populations, which suggests that colonizers were coming from multiple and genetically divergent source populations.

In a very few instances, the ages of disjunct populations are known because they were established by transplants. The insectivorous *Sarracenia purpurea* was introduced from one source in Canada into four sites in Ireland between 1963 and 1966. There were only a few founding transplants per site. Taggart, McNally, and Sharp (1990) found that the populations developing from these transplants vary substantially among each other in allele frequencies at one polymorphic allozyme locus and each differs from the source population. Most populations lack a rare allele present in the source population.

An age limit may be available for disjunct populations that occur in previously glaciated areas. Consider *Phlox pilosa* subsp. *detonsa* that occurs throughout the Coastal Plain from Tennessee to Florida and in two adjacent counties of central Illinois, roughly two hundred miles from the body of the taxon. Pollen diagrams indicate that the present vegetation of central Illinois began developing about 8,000 years ago (King, 1981). The plant's very local distribution coupled with its high abundance suggests that it has been present far less than 8,000 years. The disjunct is fixed for the predominant alleles of the taxon at two of four allozyme loci studied (Levin, 1984). It is on the verge of losing rare alleles at two other loci.

We would expect to find the amount of genetic variation in a series of islands stretching in the same general direction from the source to be related to the distance from the source. Presumably the near islands would be inhabited first and have the most variation. This scenario was explored in the rainforest tree species, *Campnosperma brevipetiolata (Anacardiaceae)*, which spread into several Pacific islands from an Indo-Malayan source (Sheely and Meagher, 1996). A trend of decreasing variation from west to east is found in the Caroline Islands, as measured by mean number of alleles per locus and mean genetic diversity per island.

If recently founded populations are small and well isolated, deleterious genes may be at unusually high frequencies by chance. This possibility is realized in *Pinus radiata*, where 3 of 15 trees that formed an isolated breeding group were heterozygous for a recessive lethal gene. An average of 6.5% of their offspring are homozygous recessive (Bannister, 1965).

Recently founded populations do not invariably have less genetic diversity than older populations, nor are they necessarily more genetically divergent. Populations may be founded by colonists from multiple sources, or younger populations may be subject to gene flow from neighboring populations. This seems to be the case in *Primula veris* (Antrobus and Lack, 1993). Some of the recent *Primula* populations are less than 100 m from source populations.

Genetic Change Facilitating Expansion

Thus far the discussion of range expansion has not included changes in the genetic composition of populations that may allow the occupation of new regions. Genetic adaptation opens invasion windows for species that lag long periods after their first introduction or which are in the process of expanding. A series of changes have occurred in *Solidago gigantea* and *S. altissima*, both of which were introduced to Europe from North America about 250 years ago. Weber and Schmid (1998) transplanted clonally propagated ramets from 24 populations of each species to a common garden at latitude 47° N. These populations ranged from Spain and Italy in the south to Sweden and Finland in the north. Both species displayed clinal variation for several traits in both species, although the slopes differed between species (figure 5.8). The level of adaptive differentiation within a relatively short time frame is quite striking. *Solidago canadensis* also was introduced into Europe from North America in the 17th century, and it too has developed latitudinal clines in some traits (Weber, 1997).

Turning to an alien in the United States, *Abutilon theophrasti* was introduced from Asia in the mid-1700s. It has been expanding vigorously northward into eastern Canada over the past 50 years. Warwick and Black (1986) grew 39 population samples from southern Ohio (39° N) to central Ontario (45° N) in a common garden. Plants from more southerly locations tended to produce a smaller number of larger seeds that were less dormant than plants from more northerly sites. Plants from more southerly locations also allocated a higher proportion of their biomass to seeds, and their seeds had a higher germination rate than plants from more northerly sites (figure 5.9). Latitudinal clines in growth form, phenology, and reproductive traits occur in other introduced species along the same gradient (*Amaranthus retroflexus*, Weaver et al., 1982; *Chenopodium album*, Warwick and Marriage, 1982a,b; *Datura stramonium*, Weaver, Warwick, and Thompson, 1985).

Divergence in flowering time also may have facilitated the recent expansion of *Trifolium subterraneum* in Australia (Cocks and Phillips, 1969), *Verbascum thapsus* in the United States (Reinartz, 1984), and *Senecio inaequidens* in Europe (Kowarik, 1995). In *Verbascum*, populations also have differentiated in habit, those in the southern part of the range being annual and those in the northern part being biennial.

Another example of divergence after long-distance dispersal is afforded by *Verbascum thapsus* (mullein), a biennial weed of temperate North America and

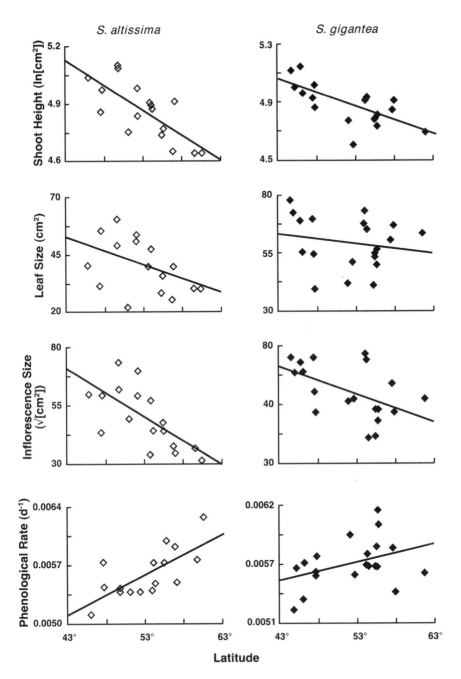

Figure 5.8. Character expressions of transplants in relation to latitude in *Solidago* species (from Weber and Schmid, 1998. Redrawn with permission of the Botanical Society of America).

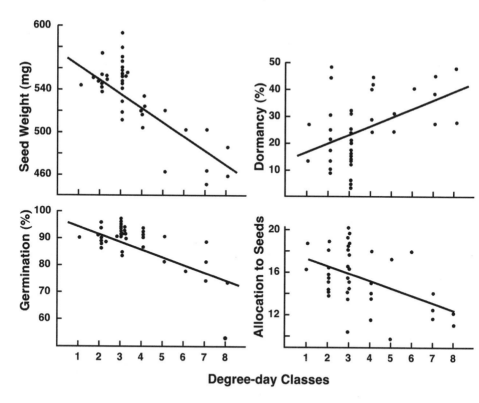

Figure 5.9. Reproductive attributes of *Abutilon* transplants in relation to degree days (from Warwick and Black, 1986. Redrawn with permission of the National Research Council of Canada).

Eurasia. Surprisingly it invaded Hawaii in the early 20th century, where it forms large populations in leeward upland (1200–3000 m) areas (Juvik and Juvik, 1992). Its success there seems to be associated with genetic changes that have led to a perennial habit, gigantism, and woodiness.

Introgression may accelerate the ongoing expansion of a species. For example, hybridization with the cultivated radish, *Raphanus sativus*, has enlarged the climatic range of the introduced weed *R. raphanistrum* in California (Panetsos and Baker, 1968).

The ability of introgression to increase the aggressive nature of species is well documented in crops, where gene receipt from wild relatives has led to the origin of weedy races. For example, a weedy race of *Pennisetum glaucum* (pearl millet) arose through hybridization with the wild *P. americanum* (Brunken, de Wet, and Harlen, 1987). Weedy sugar beets, *Beta vulgaris*, are derived from hybridization between cultivated sugar beets and conspecific wild populations (Boudry et al., 1993).

Autopolyploidy is another mechanism that affords genetic change. Successful autopolyploids usually have geographical ranges quite distinct from their diploid counterparts (W. Lewis, 1980). In *Tolmiea menziesii* (Saxifragaceae), diploid populations range from the southern limit of the species in California northward to central Oregon, where they are replaced by tetraploid populations. The latter extend northward through British Columbia into southern Alaska (figure 5.10; Soltis, 1984).

The replacement of one ploidal level by another in space or along environmental gradients is the norm (W. Lewis, 1980). This suggests that the autopolyploids often become established and spread from the periphery of the diploid's niche space. Perhaps unreduced gametes are formed more often under stressful conditions. Extremes of temperature and soil moisture deficit are the environmental influences most likely to promote the formation of unreduced gametes. The former was demonstrated in clones of *Solanum phureja* grown in a cool growth chamber (7–13° C), a warm growth chamber (12–17° C), and in the field (Veilleux and Lauer, 1981). In some clones, the percentage of unreduced gametes varies across treatments from a few percent to over 30%. There are significant genotype-environment interactions.

It follows from the foregoing that spontaneous polyploids should appear more often in geographically marginal populations of diploid species than in interior populations. Unfortunately, the literature on the occasional occurrence of polyploids in otherwise diploid species is rather sparse. There is, however, an example of spontaneous polyploids occurring in geographically marginal populations of the grass *Alopecurus bulbosus* in southern Britain (Sieber and Murray, 1980).

Whereas genetic change may facilitate the expansion of species, genetic change is not a prerequisite for species expansion. Even a reduction of genetic variation relative to the source populations does not preclude rapid range expansion. Perhaps the success is based on the use of phenotypic plasticity rather than genetic change to exploit somewhat novel environments. Niche breadth along environmental gradients does not seem to be correlated with the level of genetic variation within populations (Sultan, 1987).

The Limits to Range Expansion

Based on our understanding of the expansion of introduced weeds during the past several decades, neospecies' ranges will expand as long as there is access to habitats suitable in their biotic and abiotic characteristics (Baker, 1972; Cousins and Mortimer, 1995). Eventually all species reach the limits of suitable habitats and expansion ceases, at least for the short term.

Ecological Explanations for Range Limits

Neospecies may reach limits of suitable habitats soon after their emergence. These species will be narrow endemics. This type of distribution often arises from a very localized adaptation to narrowly circumscribed edaphic conditions such as serpentine or gypsum soils (Mason, 1946a; Kruckeberg and Rabinowitz, 1985). There are many examples of generalized species with moderate distribu-

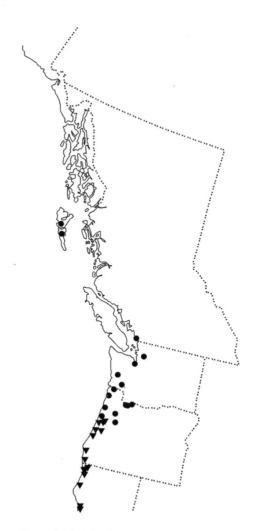

Figure 5.10. Distributions of the diploid and tetraploid cytotypes of *Tolmiea menziesii*. Triangles refer to diploid populations, and circles to tetraploid populations (from Soltis and Soltis, 1989. Redrawn with permission of *Evolution*).

tions giving rise to more specialized endemics on their geographical margin. Some genera where the pattern is common include *Linum* (Sharsmith, 1961), *Navarretia* (Polemoniaceae; Mason, 1946b); *Strepthanthus* (Brassicaceae; Kruckeberg and Rabinowitz, 1985), and *Clarkia* (Lewis, 1973).

Judging from the fact that all species once were neospecies and that most species are not edaphic specialists with narrow distributions, it follows that most neospecies are not restricted by narrowly confined soil types. Temperature, rainfall, day length, competition, herbivory, disease, the absence of pollen or seed vectors, as well as edaphic conditions, act (alone or in combination) to limit a species range. The factors that are most important in one part of the range may not be so elsewhere, and importance of any given factor may vary in time. Species often find one end of the ecological and geographical gradients along which they are distributed to be physically stressful and the other end to be biologically stressful (Brown, Stevens, and Kaufman, 1996).

The factors that limit a species' range act through their effect on plant survivorship and fecundity. The exact position of a species boundary is dictated by the interaction of birth, death, and dispersal with the spatial and temporal heterogeneity in the environment (Brown et al., 1996).

There have been many attempts to explain species margins by fitting isopleths of possible causative factors with distribution maps. For example, the restricted distributions of *Hedera helix, Tilia cordata,* and *Viscum album* in Europe can be related to specific winter minimum temperatures and summer maximum temperatures (cf. R. Crawford, 1989). However, as noted by Hengeveld (1990), the coincidence of margins and isopleths does not imply that the factors considered are necessarily causal. Most correlations between margins and isopleths for summer maximum temperature or winter minimum temperature fail to take into account the physiological tolerances of the species considered. Temperature extremes may limit species ranges through their effect on survival and/or reproduction. Pigott (1974, 1981) demonstrated that the northern limits of *Cirsium acaule* in Britain and *Tilia cordata* in Finland are due to reproductive failure. Stoller (1973) found that the northern limits of *Cyperus esculentus* and *C. rotundus* in the United States were roughly in accord with winter temperature minima. Evidence that frost sensitivity might limit their ranges was forthcoming from laboratory studies.

Margins due to abrupt edaphic shifts are the easiest to explain and to experiment with. Consider the southern margin of *Phlox drummondii* (Levin and Clay, 1984). The species occurs on loose, sandy soils in central Texas, but fails to grow on the richer, more compact soils to the south. The latter has a higher percentage of silt, clay, and organic matter, and higher levels of phosphorus, potassium, calcium, and magnesium. The transition from one soil type to another occurs within one mile. Demographic experiments were conducted in the two distinctive soil types and in the transition zone. The net reproductive rate for experimental populations established in the typical *Phlox* soil averages 14.9 versus 9.3 for populations in the transition zone and 1.3 for populations farther south. The impact of the edaphic factor without competition is evident. In the presence of competitors the net reproductive rate is 5.4 on *Phlox* soils versus 0.55 in the heavier soil. Thus populations established in the heavier soil could not replace themselves; and the species is unlikely to colonize such soil successfully.

Most experimental studies of distributional limits focus on the effects of climatic and edaphic conditions on plant performance. Carter and Prince (1981) argue that whereas these conditions may affect plant performance at existing sites, they may have an even greater role in affecting the probability that new colonies are founded.

If marginal populations could evolve, species would expand their ranges. They would become tolerant of the physical or biological exigencies that impose the limits to their present distributions. We see this at the genus level, where range expansion has occurred as a result of speciation and neospecies expansion.

Genetical Explanations for Range Limits

Ecologically and geographically marginal populations rarely deviate much from the adaptive norm characteristic of interior populations in the same area, even though these populations are exposed to long-term directional selection (Antonovics, 1976; Bradshaw, 1991). Substantial adaptive evolution appears to be the exception rather than the rule. Several genetical hypotheses have been proposed to explain this observation. Most are related either to the nature of local gene pools or to gene flow (Hoffmann and Parsons, 1997).

Bradshaw (1991) suggested that plant species do not expand their niches because they lack the genetic variation to do so. He cites many examples of populations exposed to given selective pressures that simply do not respond. The failure to do so ostensibly is not due to depauperate gene pools that may characterize small or inbred populations, but rather the limitation of genetic variation in large populations. Small effective population size would make a response to selection even less likely.

Another related, but more specific hypothesis, is that marginal populations have limited evolutionary potential because they are subjected to extreme conditions, which itself reduces heritable variation. The reduction of heritable variation when a population is exposed to greater levels of stress has been demonstrated in *Impatiens pallida* (Bennington and McGraw, 1996).

Even if marginal populations have the genetic potential to expand into new niches, they may fail to do so because alleles allowing such expansion are diluted by gene flow from interior populations. If population density declines toward species' margins, marginal populations will experience more gene flow than interior populations, unless they are well isolated by distance (Levin and Kerster, 1974). There is considerable evidence that in small populations an unusually high proportion of seeds are sired by plants outside of the population (Ellstrand, 1992a). Gene flow into marginal populations may even be maladaptive if their habitats are somewhat different from those of interior populations. As a result marginal populations may be demographic sinks (Kirkpatrick and Barton, 1997).

The aforementioned scenario is based on populations declining in size as the periphery of the range is approached. This is the case in some species growing along environmental gradients (e.g., *Vulpia ciliata*, Carey et al., 1995).

In other species (e.g., *Lactuca serriola*, Carter and Prince, 1988), population size does not decline along environmental gradients, but population longevity does. The rate of population turnover is elevated toward the periphery. In such cases, populations would be especially vulnerable to gene flow when population sizes were small. Recall that small populations always are impacted more by a given pollen or seed rain than large populations. In species with abrupt edaphic boundaries (e.g., *Phlox drummondii*, Levin and Clay, 1984), neither population size nor turnover seems to differ from the interior to the boundary. Boundary populations in these species would not have greater immigration rates than interior populations. Yet gene flow could still disrupt local adaptation.

The swamping effect of interior populations depends in part on their adaptive posture. If these populations are in the core regions of the fundamental niche space, gene flow will counter the effect of selection more than if the interior populations are toward the periphery of the niche space where the marginal populations reside.

The breeding system of populations influences their genetic variation and immigration rates, and thus has a bearing on the aforementioned hypotheses of species limits. Predominantly selfing species have less genetic variation per population than outcrossing species (Hamrick and Godt, 1996). Thus if the lack of appropriate variation is responsible for the stasis in species boundaries, selfers should be less likely to expand their boundaries than outcrossers. If gene flow is responsible for boundary stasis, then predominantly selfing species should be more likely to expand their boundaries, because they experience lower levels of immigration via pollen than outcrossing species. Most seeds are sired by the same plant or others within the same population. Moreover, advantageous genes would be more likely to be fixed in ecologically marginal selfing populations than in outcrossing populations, all else being equal. The problem, of course, is obtaining advantageous genes. Ironically, the process most likely to introduce novel genes and thus facilitate adaptation (viz. immigration) is also the process that prevents their fixation (Holt and Gomulkiewicz, 1997).

Other evolutionary hypotheses to account for limits to species ranges are discussed by Hoffmann and Blows (1994). These depend on genetic interactions among traits. One hypothesis is that range expansion requires alterations in several independent characters, so that favored genotypes are rare. Another hypothesis is that adaptation to stressful environments is limited by genetic tradeoff between fitness in favorable and unfavorable environments because the latter are not persistent. Another hypothesis is that genetic tradeoffs (or negative correlations) among fitness components in marginal environments prevent traits from evolving.

All genetical hypotheses about species range limits have predictions that are testable (Hoffmann and Parson, 1997). Unfortunately too few data are available to do this in a rigorous way. We know most about genetic variation (as judged by proportion of polymorphic loci and number of alleles at polymorphic loci) in marginal populations relative to interior ones. There is no consistent relationship between geographical position and variation (cf. Yeh and Layton, 1979).

Consequences of Ecological and Genetical Constraints

Once a neospecies has achieved its range limits, its distribution may not change substantially for some time, given that there is little change in the abiotic or biotic factors that affect its distribution. Geographical stasis requires that adaptations influencing the ecological and geographical attributes of species do not change in any significant way. Such conservatism is suggested by two lines of information. First, there is significant correlation between the range sizes (and a significant correspondence between the habitats) of disjunct herbaceous taxa within genera relict to temperate eastern Asia and eastern North America (Ricklefs and Latham, 1992). Woody plants do not show a significant spatial correlation. Woody plants may be less constrained in their ecological diversification than herbaceous plants because they tend to have more genetic diversity per population than herbaceous plants (Hamrick and Godt, 1996).

Ecological conservatism also is suggested in the numerous pollen diagrams from late-Quaternary sedimentary sequences. These show that rather than evolving to cope with climate changes in the postglacial period, nearly all Northern Hemisphere woody species migrated northward and retained their longstanding relationship with the climate (Huntley and Webb, 1989). Species migrated hundreds or thousands of miles in response to climatic change. Moreover, many ecologically and climatically similar congeners in North America and Europe have similar patterns of expansion during the Holocene, as shown for species of *Fagus* by Huntley, Bartlein, and Prentice (1989).

Given ecological conservatism and little environmental change, the distribution of neospecies usually will reflect their initial spatial relationship with their progenitors as long as the environment remains relatively constant. Neospecies that arose on the margins of their progenitor most likely will remain in spatial proximity to their progenitor for a considerable length of time. This pattern is repeated throughout flowering plants, most notably in the genus *Clarkia* (Lewis and Lewis, 1955; Lewis, 1966; Lewis, 1973). Neospecies that arose in distant disjunct populations may remain geographically remote or may approach their progenitor, depending on the adaptations of the neospecies and the character of the intervening environment. Parapatry or sympatry between these species also is possible. Thus the proximity of species' ranges is not necessarily indicative of peripatric speciation.

With ecological conservatism and considerable environmental change, both the neospecies and the progenitor are likely to track the conditions to which they are adapted by migration. Then their contemporary spatial relationship may have little information about their original spatial relationship. Neospecies that evolved on the margins of their progenitors may become geographically remote, whereas those evolving far away may become sympatric with their progenitors.

Evidence for the movement of neospecies away from their progenitors comes from studies of diploids and their allopolyploids. Allopolyploids must evolve in sympatry with both of their progenitors. Early in their histories, the distributions of most allopolyploids will be sympatric or parapatric with their progenitors. Should the climate change, the allopolyploids may become allopatric with their progenitors as a result of changes in the ranges of the progenitors or the allopolyploids (Ehrendorfer, 1980). As an example of the differential response to climate change, Stebbins (1950) refers to the *Bromus carinatus* complex, in which allopolyploids that are restricted to South America ostensibly originated via hybridization between North American diploids.

Allopolyploids also may become allopatric as a result of their dispersal. The Hawaiian endemic *Gossypium tomentosum* provides a fine example of dispersal. Its putative diploid parents have indigenous ranges in Mesoamerica and northern South America (DeJoode and Wendel, 1992). Thus dispersal to Hawaii must have occurred after hybridization and chromosome doubling and after the establishment of *G. tomentosum* in the Western Hemisphere.

Overview

The spread of neospecies is a relatively tractable subject, unlike some others being considered. We have accurate distributional records of neospecies that originated within the past few hundred years. We also can use surrogates like recent invaders to see how species spread when they encounter favorable conditions. The fossil and pollen record of plants migrating northward after the recent glacial retreat convey information on expansion over a broad time span. The surprising rate of migration owes to occasional long dispersal. As this process must be operative today, the potential for rapid expansion independent of human influence still exists. Our influence of course will accelerate the rate of expansion many fold.

Expansion is not dependent on genetic change when hospitable habitats are accessible and widespread. However, genetic change in ecologically marginal populations would allow the exploitation of habitats that otherwise would be off limits, and thus promote expansion. Gene flow from ecologically central populations tends to constrain species expansion when selection pressures are weak.

The movement of species in response to long-term climatic change appears not to require genetic change, although such movement may be accompanied by genetic change. The expansion often involves the establishment of distant foci colonized by a few individuals. This in turn may render the area invaded less variable than the source area. Reduced variation near and behind an expanding border is especially evident when one considers chloroplast haplotype diversity. The study of such diversity in herbaceous plants lags behind those on trees. Ostensibly, the northward expansion of herbs facilitated by long-distance colonization will show a signature similar to those of trees.

6

Differentiation and the Breakdown of Species Unity

An expanding neospecies moves across landscape patches to which it is adapted. Populations give rise to others that share their phenotypic and genetic attributes, thereby rendering the neospecies a unified, cohesive phenotypic entity.

The phenotypic cohesion of a neospecies initially lies in the common ancestry of its populations. Gene exchange between populations, stabilizing selection, and genetic and developmental constraints within populations restrain the neospecies from departing from its original character. Gene flow and stabilizing selection counterbalance the dispersive force of stochastic processes, whereas gene flow coupled with genetic and developmental constraints counterbalance the dispersive force of diversifying selection.

The premise of this chapter is that as a neospecies matures the phenotypic and genetic similarity among its populations—that is, the unity of the species—is likely to decline. The species disintegrates as the forces holding it together (interpopulation gene exchange, stabilizing selection, and genetic and developmental constraints) decline in strength and as the effects of diversifying selection and stochastic changes among populations increase in time. Disintegration is likely to be irrevocable.

The Differentiation of Population Systems

Adaptive differentiation has been achieved following dispersal into new habitats and following vicariant events—that is, the rupturing of species distributions by environmental forces. Both processes will be discussed in that order.

The Dynamics of Dispersal into New Habitats

The potential for the differentiation of discrete population systems following dispersal is a function of the number of populations and the number of plants in a species. The larger the distribution and size (*N*) of a neospecies, the larger the number of seeds will reach sites with somewhat different habitats. The more seeds reaching these sites, the greater is the chance of a species gaining a toehold in at least one of them.

The exploitation of new habitats is most likely if populations are founded by large numbers of immigrants. Exploitation is facilitated if the new population is largely free of gene flow from a larger population system (Paterson, 1986).

The more seeds a neospecies disperses, the more likely some will reach distant sites. Given the need for large numbers of immigrants or large pulses of immigration, it is unlikely that the invasion of very distant sites would be accompanied by ecological differentiation, because the small number of colonists probably would not contain the genetic substrate for adaptive evolution. This postulate is supported by phylogenetic studies which show that colonization following long-distance migration almost invariably involves habitats very similar to those used by the source populations (e.g., *Argyranthemum*; Francisco-Ortega et al., 1997). Ecological differentiation is more likely to accompany the invasion of not so distant sites that receive occasional, but substantial, seed rains. These sites may be either beyond or within the range of the species.

The premise that a niche shift occurs in sites somewhat removed from those occupied by "typical" populations is supported by the evolution of edaphic subspecies of *Pinus contorta* (Aitken and Libby, 1994). Subspecies *contorta* and ssp. *bolanderii* are parapatric in northern California. The former occurs on grassy coastal bluffs, the lowest (and first) of a sequence of five marine terraces. The latter occurs in a pigmy forest ecosystem that spans the four remaining terraces. The soil and species composition and balance change as one moves from the second terrace to the fifth terrace.

Allozyme data indicate that ssp. *bolanderii* is a recent derivative of ssp. *contorta*, having a subset of alleles found in the latter and only a few private alleles. What was the sequence of colonization from terrace one? If terraces two through five were simultaneously colonized, the populations on each terrace would be expected to be similarly divergent from the population on terrace one. If colonization were in a stepwise progression from terrace one to two, two to three, and so on, the populations on terrace two should be the most similar to those on terrace one, with similarity to terrace one declining with each step upward. If colonization began on a higher terrace, populations there should be the most similar to the populations on terrace one. Terrace three *bolanderii* is more similar to ssp. *contorta* than are other terraces of this subspecies, thus suggesting that colonization began at that level.

Although ecological differentiation is facilitated by geographical isolation, it is not dependent upon it (Endler, 1977). Given very strong selection, plant populations may invade neighboring habitats in the face of gene flow. This is illustrated best in the occupation of soils rich in heavy metals by several species that typically occur on normal soils (Bradshaw and McNeilly, 1981). Note that such a niche shift is predicated upon a massive seed rain from the typical habitat to the novel one. Otherwise the rare metal-tolerant genotypes would never have an opportunity to get established.

Differentiation Following Dispersal

Ecological and Geographical Races

The differentiation of populations may lead to the formation of ecological races (or ecotypes) and geographical races (or varieties or subspecies). Their distributions will be interwoven with or append the range of their progenitors. Ecological races often occupy repetitive habitat patches such as heavy-metal soils or dune soils (Grant, 1981). The same ecological race may evolve locally in several different locations, as seen in *Agrostis capillaris* (Al-hiyaly et al., 1988), *Agrostis stolonifera* (Wu et al., 1975), *Agrostis tenuis* (Nicholls and McNeilly, 1982), *Silene vulgaris* (Schat et al., 1996), and *Armeria maritima* (Vekemans and Lefebvre, 1997). The shade races of *Gilia achilleaefolia*, which appear sporadically throughout its range, also are thought to be polyphyletic (V. Grant, personal communication). Ecological races often are not morphologically well defined. There may be several ecological races in the same area as long as there is a mosaic habitat.

Geographical races occupy somewhat broader and more widely distributed habitat patches than ecological races, and are allopatric or disjunct (Grant, 1981). They sometimes are equated to climatic races when plants are distributed along elevational gradients or to subspecies. Geographical races are morphologically distinct and have different ecological tolerances.

Maritime races are distributed near inland races in several plant species. For example, the maritime race of *Layia platyglossa* occurs in isolated colonies along 300 miles of the central California coast (Clausen, 1951). The maritime race has a horizontal habit, lacks a central leader, and flowers early versus the erect habit, central leader, and later flowering of the inland race. The segregation pattern in the F_2 generation of crosses between races suggests that each of the three traits are governed by one to three genes. *Layia chrysanthemoides*, which also occurs in California, has diverged into maritime and inland races that differ from each other in much the same manner as the corresponding races of *Layia platyglossa*.

A maritime race of *Viola tricolor*, which occurs on the west Coast of Jutland, Denmark, has horizontal purplish stems, small leaves, and is perennial (Clausen, 1951). The inland form is an erect annual, with lightly pigmented stems and medium-sized leaves. The inland race has a greater penchant for selfing than the maritime race. The segregation patterns of advanced generation hybrids suggest that the aforementioned differences each are dictated by two to several genes.

Geographical races may differ in their pollination systems, as seen in *Gilia*. In *G. splendens* the widespread race has medium-long corolla tubes and is pollinated mainly by the beefly *Bombylius*. The high San Gabriel Mountain race has long corolla tubes and is pollinated by the fly *Eulonchus*. The San Bernardino Mountain race has long stout corolla tubes and is pollinated by hummingbirds; and a desert race has small dull flowers and is self-pollinating (Grant and Grant, 1965). The races replace each other in space. The development of self-pollinating races from cross-pollinating progenitors has occurred in several other *Gilia* species (Grant, 1964), presumably following the invasion of more xeric habitats and a change in pollinator mix and availability.

Parallel racial variation occurs in *Gilia* and *Eriastrum*, another member of the Polemoniaceae. In coastal California both *G. achilleaefolia* and *E. densifolium* are represented by races with broad-throated flowers in maritime areas and by small-throated races in their dry interior mountains. The maritime races of both species are pollinated by large bees, whereas the interior races of both species are pollinated by beeflies and small bees (Grant and Grant, 1965).

Differentiation following long-distance dispersal may lead to the formation of discontinuous geographical races or subspecies. This ostensibly occurred in *Nigella* in the Aegean archipelago (Strid, 1970). Whereas populations in mainland Greece gradually change with respect to several morphological traits, island populations often have sharp discontinuities. Distance is a poor predictor of the level of differentiation between islands. The outbreeding *N. arvensis* is represented by a discrete local race on nearly every Aegean island. Distinct local races of the outbreeder *N. degenii* also are found on different islands. The inbreeding *N. doerfleri* has a similar distribution to *N. degenii*, but populations on different islands are very similar to each other. Unlike the previous examples, racial differentiation in *Nigella* may in part be due to genetic drift.

Long-distance dispersal may occur not merely over tens or hundreds of miles but over thousands of miles. Raven (1963) discussed divergence in New World species with disjunctive amphitropical distributions. He cites putative introductions between North America and South America that ostensibly could not have occupied habitats in Central America and northern South America. The species are self-compatible and occur almost exclusively in open communities. The disjuncts probably were established in the late Pliocene or Pleistocene or even more recently.

Chromosomal Races

Differentiation following dispersal may lead to the formation of chromosomal races that differ either in chromosome arrangement or chromosome number (exclusive of polyploidy) from the source population. Hybrids between races are likely to have reduced fertility because of meiotic irregularities.

Parapatric chromosomal races have been described in *Gaillardia pulchella* (Asteraceae). Stoutamire (1977) described four races that radiate from Texas. One extends along the Coastal Plain to Florida and North Carolina, another to Kansas and Missouri, another to New Mexico and Arizona. The fourth is endemic to Texas. The species also has been divided into morphologically distinct varieties whose geographical distributions do not correspond to those of the chromosomal races. There are no consistent allozyme differences between the races, although differences between races are greater than differences between populations within races (Heywood and Levin, 1984).

In *Clarkia rhomboidea* there are two major chromosomal races that differ by two translocations involving three pairs of chromosomes (Mosquin, 1964). The Northern race, thought to be the ancestral one, grows in north-central California and northward to Washington and westward into Idaho and Utah. The Southern race grows in the southern Sierra Nevada, southern California, and Arizona.

In contrast to euploid or anueploid chromosomal races, some populations within species have increased their ploidal levels. *Ambrosia dumosa* offers an example of sympatric autopolyploid races (Raven et al., 1968). The diploid race occurs throughout the Sonoran Desert as does the tetraploid race. The hexaploid race is confined to California, where it is sympatric with the other races. There are no consistent morphological differences between ploidal levels.

Parapatric ploidal races are present in *Larrea divaricata* (Yang and Lowe, 1968). The diploid is narrowly distributed from southwestern Texas into southwestern New Mexico, and is replaced by the tetraploid in southern Arizona and the hexaploid in southern Nevada.

Like ecological races, autopolyploid races might have multiple origins. This is most likely in species with a penchant for producing unreduced gametes. The multiple origin of autotetraploids has been demonstrated in *Heuchera micrantha* (Soltis, Soltis, and Ness, 1989), *H. grossulariifolia* (Wolf, Soltis, and Soltis, 1990), and *Plantago media* (van Dijk and Bakx-Schotman, 1997). The spatial pattern of chloroplast DNA variation indicates that autotetraploids have arisen independently at least three times in each species at sites well removed from each other.

Differentiation Following Vicariant Events

Species are subject to major climatic and geological changes that may dissect their ranges and geographically isolate large groups of populations or population systems. Clearly the ten or more Pleistocene glacial advances and retreats in the Northern Hemisphere had a great impact on species' ranges, including the fragmentation of many. Paleoecological data offer valuable insights into species distributions and movements during the past 20,000 years (Huntley and Birks, 1983; Delcourt and Delcourt, 1987).

The Appalachian and Ozark Mountains were two prime refugial areas in the United States during the period of warming that followed the retreat of the recent Pleistocene glacier (Delcourt and Delcourt, 1987). The dissection of species' ranges set the stage for the formation of geographical races or subspecies (Mayr, 1963; Hewitt, 1989). Following amelioration of the climate, the isolated and differentiated systems expanded their ranges. For example, in the genus *Phlox*, the Ozarkian isolate of *P. divaricata* ssp. *laphamii*, migrated to the east and south while the Appalachian ssp. *divaricata* migrated primarily to the south and west (Wherry, 1955). They eventually made contact and now hybridize along a front a few hundred miles long, as will be discussed below. The subspecies differ primary in floral traits.

Asclepias tuberosa is another species in which climatic change caused the contraction of a broad range into Ozarkian and Appalachian isolates and a Florida isolate as well (Woodson, 1947). Subspeciation occurred in the isolated population systems, which expanded and met each other, forming broad zones of secondary intergradation. *Asclepias incarnata* also has subspecies that evolved in Ozarkian and Appalachian refugia, and which intergrade along their commissure.

In Europe, refugial areas for the arboreal flora include the Caucasus, northern Turkey, southern Iberia, the south Balkans, and Greece (Huntley and Birks, 1983; Taberlet et al., 1998). They were the focal points of intraspecific differentiation in many species (Ehrendorfer, 1968). In some instances differentiation occurred among disjunct remnants left behind after species had migrated northward (e.g., *Galium anisophyllum*; Ehrendorfer, 1984). In other instances, differentiation occurred after distributions were fractured by geological events (e.g., *Pinus nigra*; Ehrendorfer, 1984).

Liriodendron tulipifera provides a prime illustration of differentiation following range dissection during the Pleistocene. The species is distributed throughout most of eastern North America from south Canada to central Florida and from the Atlantic coast to the Mississippi Valley. As described by Parks et al. (1994), there are two morphologically distinctive types, one from the peninsular Florida and the other from the Appalachian highlands. The trees from Florida have more rounded leaves and smaller flowers and fruits than trees from the Appalachian region. These differences are maintained in a common garden. The genetic identity of the two groups of populations is 0.74. This value contrasts with ca. 0.92 for populations in the same region.

The expected concordance between the approximate age of geographical isolation and allozyme divergence occurs in *Wislizenia refracta* (Capparidaceae; Vanderpool, Elisens, and Estes, 1991). This species is composed of three allopatric subspecies: ssp. *californica*, a summer annual restricted to the central valley of California; ssp. *refracta*, an annual that occurs in the Great Basin, and in the Mojave, Sonoran, and Chihuahuan deserts; ssp. *palmeri*, a perennial of lower elevations in the Sonoran Desert of southern California, Baja California, and Sonora.

The genetic identities between ssp. *refracta* and ssp. *californica*, on the one hand, and ssp. *palmeri*, on the other, are 0.85 and 0.70, respectively. The identity of ssp. *palmeri* and *ssp. californica* is 0.76. Subspecies *palmeri* and ssp. *refracta* occur on opposite sides of the Sierra Madre Occidental in Mexico (early Tertiary orogeny), and may have been isolated by the rise of this mountain range. Based on allozyme frequency differences, these subspecies are thought to have diverged from 3.2 to 5.0 Myr before present. Divergence between ssp. *refracta* and ssp. *californica* is thought to have occurred from 0.79 to 0.85 Myr before present, which is in accord with a hypothesized late Pleistocene origin of the Central Valley of California (Raven and Axelrod, 1978).

Genetic Identities of Intraspecific Taxa

The genetic identities of intraspecific taxa are related to their geographical relationships (table 6.1). The taxa with spatial proximity tend to be more similar than distant taxa. Infraspecific taxa in several species have genetic identities of 0.99. In all instances, these elements are in close proximity. Neighboring taxa within *Coreopsis grandiflora* have mean identities of 0.93. Although lower than in other species, the identity of taxa is similar to identities of populations within taxa. In several species, mean identities of elements are less than 0.90. These taxa have substantial geographical gaps between them in nearly all instances.

The genus *Senecio* has a particularly interesting set of geography:identity relationships (Liston, Rieseberg, and Elias, 1989). Two widely disjunct subspecies of *S. flavus* in North Africa have an I value of 0.85. Disjunct population

Table 6.1. The mean genetic identities of proximal and remote intraspecific taxa.

Species	I	Reference
PROXIMAL RELATIONSHIP		
Phlox drummondii	0.99	Levin, 1977
Plantago major	0.99	van Dijk and van Delden, 1981
Limnanthes alba	0.99	McNeill and Jain, 1983
Astragalus tener	0.99	Liston, 1992
Astragalus rattani	0.99	Liston, 1992
Gaillardia pulchella	0.95	Heywood and Levin, 1994
Cypripedium calceolus	0.95	Case, 1993
Allium douglasii	0.93	Rieseberg et al., 1987
Coreopsis grandiflora	0.93	Crawford and Smith, 1984
Sullivantia hapemanii	0.93	Soltis, 1981
REMOTE RELATIONSHIP		
Chenopodium incanum	0.94	Crawford, 1977, 1979
Senecio flavus	0.85	Liston et al., 1989
Campanula punctata	0.84	Inoue and Kawahara, 1990
Solanum heterodoxum	0.83	Whalen, 1979
Allium douglasii	0.75	Rieseberg et al., 1987
Coreopsis cyclocarpa	0.75	Crawford and Bayer, 1981
Limnanthes floccosa	0.74	NcNeill and Jain, 1983
Solanum citrullifolium	0.71	Whalen, 1979
Styrax officinalis	0.58	Fritsch, 1996

assemblages of *S. flavus* ssp. *flavus* in North Africa and Namibia have approximately the same value. In contrast, *S. flavus* ssp. *breviflorus* and *S. mohavensis* of the American Mojave Desert have a genetic identity of 0.95. Ostensibly, the intercontinental disjunction giving rise to a new species occurred subsequent to the African disjunctions.

Disjunction associated with the colonization of a distant island is known in *Campanula punctata* (Inoue and Kawahara, 1990). The mean genetic identity of populations on mainland Japan and on the Izu Islands is 0.84. Populations on more distant islands tend to have lower identity with the mainland than populations on islands close to the mainland.

If a novel population system arose in recent times, its genetic identity with the progenitor should be high. This premise may be tested either in wild taxa that evolved in areas liberated by the recent glacial retreat or in cultivars. The derivation of cultivars from their wild ancestors is a process similar to divergence following dispersal, in that a genetic bottleneck is coincident with strong selection on one or more traits.

Consider first a recently evolved subspecies of *Phlox pilosa* that is an endemic in central Illinois. This subspecies (*sangamonensis*) and its putative ancestor subspecies *detonsa*, which occurs throughout much of the Coastal Plain in the southeastern United States, have a genetic identity of 0.97 (Levin, 1984). The time of disjunction is likely to be less than 4,000 years.

It is common for cultivars to differ little from their wild ancestors in allozyme constitution, even though they differ substantially in a few phenotypic traits. For example, the genetic identity of wild and cultivated *Capsicum baccatum* is 0.98 (McLeod et al., 1983). The mean genetic identity of wild *Helianthus annuus* and Native American cultivars is 0.95; the more highly derived cultivars have a mean I value of 0.93 (Cronn et al., 1997). This compares to identities of 0.96 between widely spaced populations of *H. annuus*.

Hybrid Zones

Intraspecific taxa that initially are allopatric may come into contact and form long-lived hybrid zones. How these zones are maintained depends on (1) the crossability of the taxa and the fitness of their hybrids, and (2) the ecological disparity of the taxa (Barton and Hewitt, 1985; Hewitt, 1988). Zones may be maintained if taxa have similar habitat requirements, but hybrids are less fit than their parents. This is referred to as the dynamic equilibrium model or hybrid disadvantage model (Barton and Hewitt, 1985; Moore and Price, 1993). The width of the zone is determined by the balance between dispersal and selection against hybrids. The broader is dispersal or the weaker is selection, the wider will be the zone. Long-distance dispersal and the interdigitation of pure populations prior to contact will result in the widest zones (Hewitt, 1993).

The hybrid zone may lie along an environmental transition with one taxon being adapted to conditions at one end and the second taxon to conditions at the other end. The zone will be maintained by selection against parental and hybrid types in alien environments. If hybrids are superior to the parental taxa in the intermediate environment, this is referred to as the bounded hybrid superiority model or hybrid advantage model (Endler, 1977; Moore and Price, 1993). If hybrids were not superior, this is referred to as the environmental cline model. The width of the cline is dictated by the slope of the environmental gradient. Clines among several traits need not be coincident. Those traits (genes) subject to selection should have steeper clines than traits that were neutral. Included in the latter would be cytoplasmic markers.

Neutral Hybrid Zones

Hybrid zones may involve taxa with substantial genome compatibility and similar habitat requirements. In this case, the zone will not maintain its width. Neutral diffusion produces zones with concordant clines, which over time become broader and shallower (Hewitt, 1988). The zone may not maintain its position either. If one taxon is more abundant or has greater dispersability, the zone will be pushed in the direction that the taxon was migrating.

During the past few thousand years, *Phlox divaricata* ssp. *laphamii* expanded from the Ozarks into Illinois, Kentucky, and Tennessee, where it met ssp. *divaricata* coming out the Appalachians. A hybrid zone extends from the southern tip of Lake Michigan southward into Kentucky and southeastward into central Tennessee (Levin, 1967). As judged from morphological traits, the zone is about 150 miles wide, with a weakly irregular transition from one subspecies to the other. There are no pre- or postzygotic barriers to hamper gene flow between them. Thus it appears that this hybrid zone is expanding by neutral diffusion.

The outcrossing, annual *Phlox drummondii* provides another example of a hybrid zone generated by neutral diffusion. In this case there is experimental evidence that selection is not a significant factor. The red-flowered ssp. *drummondii* and pink-flowered ssp. *mcallisteri* intergrade across a two-mile-wide belt in central Texas (Levin and Schmidt, 1985). The success of parental and hybrid subpopulations in both parental habitats and in zone habitats were assessed in a reciprocal sowing program. Net reproductive rates of both subspecies and hybrids are similar across sites, indicating the absence of exogenous selection on aliens or hybrids. There are no intrinsic barriers to subspecific gene exchange, and pollen and seed dispersal are restricted (Levin, 1983b). The narrow width of the zone ostensibly reflects its recent origin.

If a hybrid zone is the product of neutral diffusion, the width of the zone will gradually expand, at a rate dependent on gene dispersion (Endler, 1977; Barton and Hewitt, 1985). If two taxa meet along a uniform front, the width of a cline (w) is expressed as a function of time since neutral secondary contact, as follows:

$$w = 1.68 \, l \, (\sqrt{T})$$

where l is the dispersal distance per generation and T is time measured in generations (Endler, 1977). If the dispersal distance were 100 meters per generation and 1,000 generations had passed, the zone would be about 5 km wide. If the taxa interdigitated along the front or if the front initially was formed by a mosaic of the taxa—that is, some populations of taxon A were established beyond the border of taxon B and vice versa—the width of the zone would be much greater at time T, all else being equal.

The problem of studying a hybrid zone is to observe a static structure on the one hand, and see evidence of a change in time on the other. This can be accomplished when we compare gene frequencies in adults with gene frequencies in their juveniles. If a hybrid zone is expanding, the level of alien genes (or admixture) in juvenile subpopulations should be greater than in adult subpopulations at the same localities. This is the case in *Pinus muricata* adjacent to a narrow hybrid zone (Millar, 1983). In northern California two population systems differing in morphology, terpene chemistry, flowering time, and allozyme frequency are separated by a zone less than 3 km wide. The fossil record indicates that contact between the two population systems is recent. Allele frequencies at the GOT-1 locus, which are very different between the systems, differ between embryos and adults (figure 6.1).

Zones Maintained by Habitat Differences

Hybrid zones may be maintained through different habitat preferences without significant genome incompatibility, as described in *Artemisia tridentata* (Wang et al., 1997). Subspecies *tridentata* and ssp. *vaseyana* are parapatric and segregate along strong elevational and topographic gradients. In northern Utah ssp. *tridentata* typically grows below 1800 m in elevation, while ssp. *vaseyana* typically grows above 1900 m. A hybrid zone occurs between these elevations. The zone probably was formed since the last glaciation.

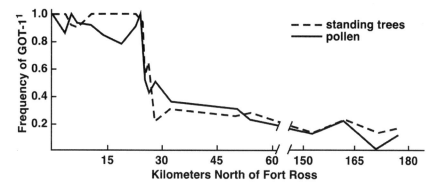

Figure 6.1. Cline in the frequency of *Got-1¹* in bishop pine (from Millar, 1983. Redrawn with permission of *Evolution*).

Reciprocal transplant experiments demonstrate that plant performance is dependent on the source of plants and where they are planted. When the parental subspecies are planted in each other's habitats, the native subspecies always has higher fitness than the alien subspecies, as measured by plant survivorship and inflorescence size. Within the hybrid zone, native hybrids have greater fitness than either of the subspecies or hybrids from other parts of the zone.

Nagy and Rice (1997) conducted a reciprocal sowing experiment on two races of the annual *Gilia capitata* with disparate habitat preferences. Seeds were planted in and exchanged between an inland site and a coastal site in California where the species grows. They estimated net reproductive rates from demographic data. In 1993, natives at the inland site had a threefold advantage over aliens in R_o, whereas at the coastal site they had a hundredfold advantage.

Nagy (1997a) modeled the evolutionary consequences of hybridization between the inland and coastal populations by generating F_2 seeds and planting them in each site. He then measured phenotypic selection on each population using a multiple regression analysis on leaf shape, petal shape, and petal color. Selection at both sites favors the inland leaf morphology. The inland petal shape and color are favored at the inland site, but neither is correlated with fitness at the coastal site.

Phenotypic selection may not lead to an evolutionary response, and if it does the response may not mirror the direction or intensity of phenotypic selection. In order to measure the response to selection in *Gilia capitata*, Nagy (1997a) grew the progeny from the F_2 plants of the coastal and inland sites in the greenhouse and tested for differences in trait expression. Leaf shapes of the F_3 offspring of individuals selected at the coastal site differed significantly from those selected at the inland site. Petal shape and color also differed significantly between the two groups of F_3s. Natural selection had shifted both hybrid populations toward the native phenotypes at both sites. However, selection against alien traits was stronger at the inland site. Thus experimental gene flow had been counterbalanced by selection.

Zones Maintained by Habitat Differences
and Genome Incompatibility

Both alien disadvantage and genome incompatibility maintain a hybrid zone between two varieties of *Gaillardia pulchella* in central Texas. One variety occurs on soils low in clay content and high in carbonate, and the other on soils high in clay and low in carbonate (Heywood, 1986a). The transition from one soil type to another occurs within 200 m or less. The two varieties have very different frequencies of *Me* alleles at 2 loci in addition to different patterns of ray floret pigmentation. The frequencies of the *Me* alleles change sharply in accord with the soil transition at each of three transects, generating clines a few hundred meters wide or less (figure 6.2). The shift in ray pigmentation occurs over a longer distance (2–3 times greater than for *Me*).

It is not unusual for unrelated traits in an outbreeding species to display clines of different slopes (Endler, 1977; Barton and Hewitt, 1985). Traits more closely associated with adaptation better track changes in key environmental variables. In predominantly inbreeding species, a greater concordance in clines might be expected even if some traits were unrelated to fitness because of genetic hitchhiking (Hedrick and Holden, 1979). Inbreeding species also would be expected to have narrower clines than outbreeders, all else being equal, because self-fertilization reduces gene flow within and between plant types.

Intraspecific population systems that differ in ploidal level also make contact. Hybrids typically are sterile, so backcrossing is largely precluded. Such contact zones almost invariably are maintained by disparate habitat preferences as well as genome incompatibility. One example is in *Chamaenerion angustifolium* (Onagraceae), where diploids and tetraploids are geographically separated by latitude (Husband and Schemske, 1998). Tetraploids occur at lower latitudes. Contact between these cytotypes extends across North America near the southern limit of the boreal forest and also extends southward along the Rocky Mountains, with tetraploids occurring at the lower elevations.

Husband and Schemske (1998) determined the frequency of diploids, triploids, and tetraploids along 27 transects near the Montana-Wyoming border. Most populations contain only diploids or only tetraploids; others contain predominantly one cytotype or the other. Some populations contain nearly similar proportions of the two cytotypes. Mixed populations may or may not have triploids. When present they occur with lower frequencies than expected by random mating due to pre- and postzygotic barriers. The cytotypes did not gradually replace each other in space; rather they are distributed in a mosaic pattern. This is evident within (figure 6.3) and among transects.

The aforementioned example is unusual in that zones of contact between population systems with different ploidal levels rarely contain hybrids. Some well-studied examples of contact zones include *Tolmiea menziesii* (Soltis, 1984), *Plantago media* (Van Dijk, Hartog, and Van Delden, 1992), *Deschampsia caespitosa* (Rothera and Davy, 1986), and *Dactylis glomerata* (Lumerat et al., 1987). Differences in phenology and crossing barriers between cytotypes are the prime reasons for the lack or paucity of hybrids.

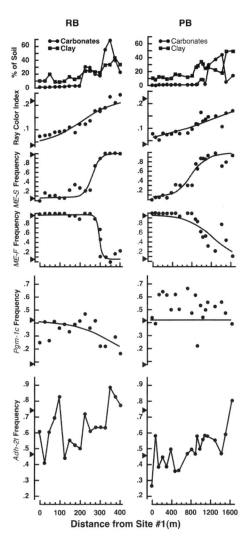

Figure 6.2. Transects through population RB (left) and population PB (right) of *Gaillardia pulchella* (from Heywood, 1986. Redrawn with permission of *Evolution*).

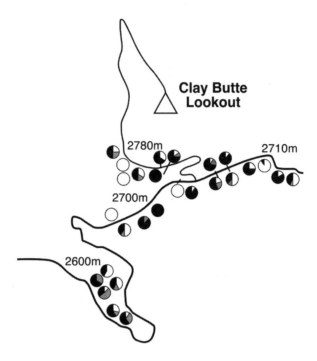

Figure 6.3. The frequency of diploids (white), triploids (grey), and tetraploids (black) in populations of *Chamaenerion angustifolium* near the Beartooth Highway (from Husband and Schemske, 1998. Redrawn with permission of the Botanical Society of America).

Contact zones of two ploidal races are likely to be parapatric and remain so if the races have the same ecological tolerances. Intrusion of one cytotype into populations of the other will be opposed, because the immigrant cytotype will be at a mating disadvantage and hybrids will be sterile (Lewis, 1967; Levin, 1975). If hybrids are not sterile, limited introgression may occur primarily in the direction of the higher ploidal races, as in *Dactylis glomerata* (Lumaret et al., 1989), *Zea* (Doebley, 1989), and *Plantago media* (Van Dijk and Bakx-Schotman, 1997). If the ploidal races have dissimilar ecological tolerances, the contact zone initially may be a mosaic, and later expand into a zone of sympatry.

Asymmetrical Hybrid Zones

The hybrid zones discussed above are symmetrical—that is, the impact of hybridization is roughly equal on both species. The zone centers along the line of species contact and gene frequency gradients from the center of the zone outward are similar in both directions. Not all hybrid zones, however, are symmetrical (Barton and Hewitt, 1985). Differing selection intensities on hybrids in parental environments or reciprocal differences in crossability or hybrid fitness can produce asymmetry. So can intertaxon differences in pollen or seed dispersal distances and differences in population size, especially in the case of neutral diffusion.

The best examples of asymmetry involve interspecific hybrid zones. Keim et al. (1989) and Paige, Capman, and Jennetten (1991) describe a hybrid zone between *Populus fremontii* and *P. angustifolia* in which there is an asymmetrical distribution of nuclear and cytoplasmic genotypes; *P. angustifolia* is the primary recipient of gene flow. This is consistent with observations that backcrosses involving *P. fremontii* are inviable, whereas those involving *P. angustifolia* are healthy.

The second example of an asymmetrical hybrid zone involves differential pollen dispersal. *Aesculus pavia* and *A. sylvatica* meet near the Fall Line, which marks the boundary of the Piedmont and the Coastal Plain. Most of the hybrid zone lies on the *A. sylvatica* side of species contact (dePamphlis and Wyatt, 1990). Hybrids are present at least 150 km beyond the range of *A. pavia*, but rarely more than 25 km beyond the range of *A. sylvatica* (figure 6.4). This is attributed to the activities of migratory hummingbirds, which move northward in the spring carrying pollen of *A. pavia* into the range of *A. sylvatica*.

If two taxa do not have disparate ecological tolerances, their hybrid zones will not necessarily be centered on an environmental transition or discontinuity. Then a selective advantage of one taxon over another or asymmetrical gene flow can move the hybrid zone in the direction of the inferior taxon. This may be happening in the *Populus* example cited above and in a hybrid zone involving *Ipomopsis aggregata* and *I. tenuituba* (Campbell, Waser, and Melendez-Acker-

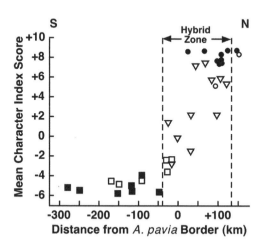

Figure 6.4. Variation in mean index scores for populations in a hybrid zone between *Aesculus pavia* and *A. sylvatica* ■ *A. pavia*; □ *A. pavia* with hybrids; ▽ hybrid zone populations; ○ *A. sylvatica* with hybrids; ● *A. sylvatica* (from Depamphlis and Wyatt, 1990. Redrawn with permission of *Evolution*).

man, 1997). Hummingbirds favor traits of the former (wide corolla tubes and intense red pigmentation), whereas hawkmoths favor traits of the latter (narrow corolla tubes and pale or no pigmentation). In the hybrid zone, hummingbirds typically are the predominant pollinator, and thus favor traits of *I. aggregata*.

The Breakdown of Species Cohesion

Interpopulation divergence through selective or stochastic processes reduces the phenotypic unity of species. It also reduces the genetic unity of species by reducing the ability of populations to exchange genes by pollen or seeds.

Reproductive Isolation

Ecological differentiation within a species may or may not be accompanied by the emergence of postpollination barriers to gene exchange. The races of *Potentilla glandulosa*, *Layia platyglossa*, and *Gilia achilleaefolia* referred to above are fully intercrossable and hybrids have close to normal fertility. Weak barriers between races is the norm, unless chromosomal changes are involved (Grant, 1981). Some exceptions to the rule are discussed below.

The incidental development of cross-compatibility barriers between populations has ensued in many genera including *Gilia* (Grant, 1952, 1954), *Strepthanthus* (Kruckeberg, 1957), *Solanum* (Grun, 1961), *Lycopersicon* (Rick, 1963), and *Mimulus* (Vickery, 1978). Cross-incompatibility in these studies is expressed as reduced seed-set. The basis for reduced seed set was not investigated.

Cross-incompatibility in *Phlox drummondii* was measured in terms of the percentage of pollen grains sending tubes into the style. A lower percentage of pollen germinated when ecologically and geographically distinct subspecies are crossed than when populations within subspecies are crossed (Levin, 1976b). More ecologically divergent subspecies have about 25% reduction in pollen tube germination versus about a 10% reduction in less divergent subspecies.

The differential success of germinating pollen grains from different populations is illustrated in *Turnera ulmifolia*. Baker and Shore (1995) performed mixed pollinations where self pollen competed with cross pollen from the same population or pollen from other populations. The performance of self pollen relative to extraneous pollen was very much better than it was relative to cross pollen from the same population. Instead of siring about 50% of the progeny from 1 : 1 pollen mixtures, self pollen sired over 90% of the progeny in some combinations of populations.

The genes responsible for a shift in ecological preference may themselves produce postpollination barriers to gene exchange between ecologically divergent populations. This possibility is most striking in *Mimulus guttatus* (Macnair and Christie, 1983). Crosses between a copper-tolerant population and a nontolerant population yield F_1 hybrids, many of which died as juveniles. A gene pleiotropic with or tightly linked to copper tolerance may be responsible for this partial barrier to gene exchange.

Reduced hybrid fertility may be a by-product of race formation (Levin, 1978a, Grant, 1981). The reduction may be only a few percent relative to crosses within a taxon (e.g., *Silene*, Kruckeberg, 1961; *Lasthenia*, Ornduff, 1966) or it may be more than 50% (e.g., *Helianthus*, Heiser, 1961; *Clarkia*, Vasek, 1964). Large reductions almost invariably are associated with chromosomal divergence.

The fertility of hybrids has been assessed in relation to the distance between populations in several species. In some species fertility decreases as the distance between populations increases, as in the *Strepthanthus glandulosus* complex (Kruckeberg, 1957). This relationship is depicted for *S. glandulosus* ssp. *secunda* in figure 6.5. Distance-dependent fertility also has been described in the *Gilia achilleaefolia* complex (Grant, 1954), the *Luzula campestris-multiflora* complex (Nordenskiold, 1971), and in *Epilobium*, Raven and Raven, 1976). In other species, there is no relationship with distance (e.g., *Atriplex triangulata* complex, Gustafsson, 1974; *Clarkia rubicunda*, Bartholomew, Eaton, and Raven, 1973).

The likelihood of novel chromosomal arrangements being fixed within populations is an inverse function of population size and a positive function of geographical isolation because stochastic processes are involved. Therefore, we would expect hybrids between small populations to have lower fertility on average than hybrids between large populations. This expectation is realized in Scandinavian populations of *Atriplex longipes* (figure 6.6; Gustafsson, 1974).

Given that postpollination barriers are by-products of divergence, we might expect populations separated by long spans of time to have strong barriers.

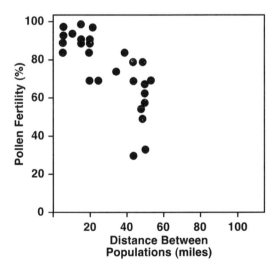

Figure 6.5. The relationship between pollen fertility in hybrids and interparent distance in *Strepthanthus glandulosus* ssp. *secunda* (from Kruckeberg, 1957. Redrawn with permission of *Evolution*).

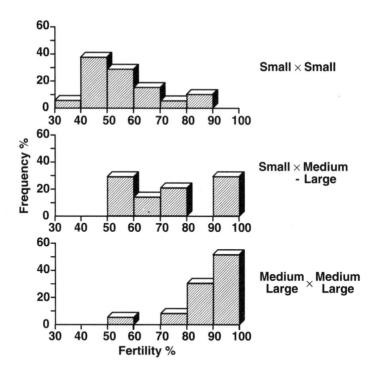

Figure 6.6. The distribution of mean fertility values in crosses between *Atriplex longipes* populations in relation to population size (from Gustafsson, 1974).

This is not necessarily the case. This can be illustrated with ancient disjunct species. *Liriodendron tulipifera* of eastern North American and *L. chinense* of eastern Asia have a genetic identity of 0.43 and a chloroplast DNA sequence divergence of 1.24% (Parks and Wendel, 1990). These molecular data sets and paleobotanical data concur in suggesting a divergence time of 10 to 16 million years before present. The species hybridize readily and produce heterotic, semisterile F_1 hybrids and vigorous F_2s (Parks et al., 1983).

Other intercrossable ancient disjuncts occur in *Liquidambar*. The eastern North American *L. styraciflua* and the east Asian *L. formosana* have a genetic identity of 0.48, and may have diverged as much as 16 million years ago (Hoey and Parks, 1991). Hybrids have been obtained, but they are not fertile. Little morphological and ecological divergence has occurred in *Liquidambar* and *Liriodendron* in spite of the great allozyme divergence.

Character Displacement

The presence of a related species may constitute an important stimulus for the divergence of local populations from the remainder of the species. Divergence in floral traits and phenology may reduce gametic wastage and intensify barriers to gene flow. Divergence in habitat requirements may relieve close competition and permit the contestants to exploit their environments more efficiently. Character divergence in the face of some challenge from a related species has been referred to as "character displacement" (Brown and Wilson, 1956). The term "Wallace effect" has been applied specifically to responses that reinforce previously existing reproductive barriers (Grant, 1966a).

The sole criterion of character displacement in many plant and animal studies has been a difference in the character states of a species in sympatry and allopatry with a related species. This criterion is insufficient, because it does not demonstrate that the related species was the stimulus for divergence (P. Grant, 1972). The detection of character displacement requires that the following questions be answered: What was the pre-contact character state? Are the differences in pre- and post-contact character states due to selection arising from the presence of the related species? Shifts in character state could be due to (1) selection to reinforce some difference, (2) selection on the character having nothing to do with the presence of the congener, or (3) a correlated response to selection on another trait (Waser, 1983).

The most convincing demonstration of character displacement in plants involves *Phlox drummondii* (Levin, 1985). This species has pink corollas throughout much of its range, which is confined to Texas. Where it is sympatric with the pink-flowered *P. cuspidata*, it has red corollas. The red-flowered *P. drummondii* is the only red-flowered *Phlox*. Thus red must be a derived state.

Phlox drummondii and *P. cuspidata* are serviced by a similar array of lepidopterans and hybridize in nature. The phenological and crossing barriers between them are modest. What is the advantage of red flowers for *P. drummondii* in the presence of *P. cuspidata*? To answer this question, Levin (1985) introduced an equal number of red- and pink-flowered plants of *P. drummondii* into a large population of *P. cuspidata*. Thirty-eight percent of the seed from pink-flowered plants were hybrids, whereas only 13% of the seed from red-flowered plants were hybrids.

Hybridization with *P. cuspidata* depresses conspecific seed production in *P. drummondii*. Alien pollen receipt is likely to be accompanied by pollen loss to the alien. Accordingly any floral variant that reduces hybrid seed production and pollen wastage has a selective premium. Since hybridization with red-flowered plants is much less than with the pink, and since pollen wastage ostensibly is much less as well, it is likely that red flowers are a consequence of natural selection for increased reproductive isolation.

Reproductive character displacement for flower color also has been documented in areas of local sympatry between *Phlox pilosa* and *P. glaberrima*. Both species have pink corollas, except in several areas of contact wherein a white variant prevails in *P. pilosa*. The white variant normally is very rare or absent. In mixed populations with *P. glaberrima*, the white variant receives less alien pollen per stigma than the pink (Levin and Kerster, 1967) and produces a lower percentage of hybrid seed (Levin and Schaal, 1970).

Experimental evidence of character displacement is lacking in other genera. Differences in floral color between sympatric and allopatric populations of species in *Clarkia, Fuchsia,* and *Rudbeckia,* and differences in floral form in

Table 6.2. Percentage hybrids produced from pollen mixtures in corn (after Paterniani, 1969).

	Pollen Source			
Female Parent	Flint orig. + Sweet orig.	Flint orig. + Sweet IV	Flint IV + Sweet orig.	Flint IV + Sweet IV
Yellow Sweet original	50.1	48.8	45.6	41.6
Yellow Sweet IV	52.0	49.3	54.7	44.7
White Flint original	45.5	51.8	30.8	61.5
White Flint IV	19.4	27.1	21.8	29.3

Polanisia and *Solanum* could be the result of selection for enhanced reproductive isolation, as discussed by Rathcke (1983).

The incidence of hybridization has been modified through artificial selection. One such program involved selection to increase cross-incompatibility between *Phlox drummondii* and *P. cuspidata* (Fritz, 1998). These species cross readily in the greenhouse and produce natural hybrids. After two generations of selection for cross-incompatibility, *P. drummondii* produced 0.58 seeds per ovule with pollen from *P. cuspidata* as compared to 0.79 seeds per ovule in the control population. The difference is statistically significant. Selection on *P. cuspidata* failed to yield a significant response.

Selection against hybridization between white flint maize and yellow sweet maize was conducted in field plots (Paterniani, 1969). After five generations of selection, the percentage of hybrids produced by white flint declined from 36% in the original population to 5%, and the percentage produced by yellow sweet declined from 47% to 3%. The strains diverged in flowering time, although they had similar phenologies to begin with. The selected white flint began to flower five days earlier than the original population and the selected yellow sweet two days later.

Paterniani also found that a crossing barrier had arisen in the white flint selection line. In the original populations, 1:1 pollen mixtures of the two strains, produced close to 50% hybrids in both strains. This is to be expected with complete cross-compatibility. However, when white flint of the fourth selection generation received 1:1 pollen mixtures of various types, less than 30% of the progeny were hybrids (table 6.2). Yellow sweet pollen was at a competitive disadvantage to white flint pollen in white flint stigmas and styles. Oddly enough the selected yellow sweet line did not become more discriminating when used as the female parent.

Overview

Given the local origin of species, it follows that populations initially will be similar in their ecological strategies and will be genomically compatible. They will remain so as long as selective and stochastic processes do not fracture this unity. The broader the species' distribution and the older the species, the greater the likelihood of disintegration, especially if the species distribution is discontinuous.

We know more about intraspecific differentiation than any other aspect of a species' passage. Detailed spatial patterns of phenotypic variation and ecological tolerances have been reported for well over a half century; and the number of demonstrations continues to grow, with emphasis on a very local scale. Over the past three decades attention has turned to the spatial patterns of molecular markers, which for the most part, are not shaped by selection. These patterns are not necessary congruent with phenotypic or tolerance patterns, nor are they congruent among themselves. The use of molecular markers has made us much more adept at recognizing the effects of stochastic processes.

Until the past decade, there were few studies of plant hybrid zones, although it has been a popular topic in zoological circles for the past 30 years. Now molecular markers are used to describe these zones in plants, and demographic studies are employed to explain their dynamics.

Many zoologists have argued that character displacement may magnify differences between closely related taxa. Although there are many demonstrations of such in the animal literature, there are very few in the plant literature. In plants, the presence of a related taxon is much more likely to result in hybridization than in character displacement.

7

The Decline and Demise of Species

At some point in their history, species face a deteriorating environment across their range. Such deterioration may place a species on a trajectory toward rarity and ultimately extinction. Schonewald-Cox and Buechner (1991) proposed four possible forms this trajectory may follow (figure 7.1).

1. The number of individuals declines but the range size remains constant.
2. The number of individuals declines in concert with range size such that the density of individuals remains constant.
3. The number of individuals and range size decline in concert, but the number of individuals declines at a faster rate than range size, leading to a reduction in the density of individuals over time.
4. The number of individuals and range size decline in concert, but the number of individuals declines at a slower rate than range size, leading to an increase in the density of individuals over time.

As discussed by Gaston (1994), trajectory (1) seems unlikely because the collapse in total population size does not impact range size. Trajectory (2) may be common when species face continuous habitat loss, but where persisting habitat patches retain their original character. Trajectory (3) may reflect range erosion due to habitat loss coupled with the degradation of the remaining habitat. Trajectory (4) may result when those areas in which a species achieves higher densities are differentially preserved.

Given the feasibility of trajectories 2 to 4, the one a given species is likely to take is impossible to predict, unless we have very specific information on the long-term causes of decline and their distribution in space. The better we understand the environmental conditions correlated with the presence of a species, the more informed our speculations will be on the consequences of

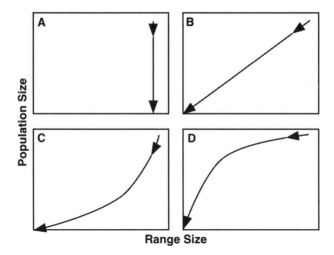

Figure 7.1. Models of the trajectories of abundance and range size towards extinction (from Schonewald-Cox and Buechner, 1991. Redrawn with permission of Birkhäuser Verlag AG).

environmental change. Whether we have such understanding can be tested on contemporary species distributions as Margules and Austin (1995) have done on *Eucalyptus radiata* in New South Wales, Australia. They calculated the probability of the species occurrence at all combinations of rainfall, temperature, and rock types in their study area. They generated a geographical map of the probability of species occurrence for each 1/100-degree grid cell. There was a good correspondence between the actual and predicted local distribution.

Kinds of Rarity

Contemporary rare species would seem to be a fertile field in which to interpret trajectories of species decline. However, rarity does not necessarily signal a species in decline (Kruckeberg and Rabinowitz, 1985; Fiedler and Ahouse, 1992). Some species may be recent endemics (e.g., *Clarkia biloba, Limnanthes bakeri*); others may be ancient endemics (e.g., *Torreya californica, Ginkgo biloba*). Unfortunately, we know little about the abundance and distributions of rare species over long periods of time.

Rabinowitz (1981) described alternate forms of rarity on the basis of the geographical distribution (wide, narrow), habitat specificity (broad and narrow), and local population size (everywhere large and everywhere small). All combinations of distribution, habitat specificity, and population size (except narrow range, small populations, and broad habitat specificity) are feasible, thus yielding seven forms of rarity. The distribution of 160 rare British species among these forms is shown in table 7.1 (Rabinowitz, Cairns, and Dillon, 1986). Most rare species have wide distributions and nearly all have large populations somewhere. Fiedler and Ahouse (1992) categorize rarity within a temporal as well as spatial context. Temporal persistence may be recognized as long or short and spatial distribution as wide or narrow. Rare species in the short/wide category never are locally abundant. Many are herbaceous neoendemics occurring in species-rich environments (e.g., the north American orchid *Isotria medeoloides*). Species in the long/wide category are not locally abundant. Most are trees or shrubs of species-rich environments. Some are paleoendemics (e.g., the California nutmeg, *Torreya californica*). Many species in the short/narrow category are locally abundant (e.g., the California vernal species *Limnanthes bakeri*). Species in the long/narrow category vary in local abundance. Typically they are trees or shrubs and are members of a pre-Pleistocene flora (e.g., the Macnab cypress, *Cupressus macnabiana*).

Temporal Changes in Abundance

The temporal history of species provides some insights into the status of species rarity. Neoendemics may not be in a state of decline, especially if they are adapted to specialized habitats and are locally abundant. Some paleoendemics may be in a state of decline.

In a few instances we know that a locally restricted species was formerly widespread and common. For example, the giant sequoia (*Sequoiadendron giganteum*) and the California redwood (*Sequoia sempervirens*) are now restricted to unglaciated valleys of the Sierra Nevada and relictual habitats along the fog belt of the California coast, respectively, whereas both were very widely distributed before the Pleistocene glaciations (Fiedler and Ahouse, 1992). The California fan palm (*Washingtonia filifera*), which appears to be a relictual endemic in the southwestern United States and Baja California, Mexico, is another example of a species that once had a much broader distribution. This species ostensibly is rare because of the severe and rapid climatic changes that ensued since the Miocene. It is restricted to oases associated with the San Andreas Fault. Another example is the Chihuahuan spruce, *Picea chihuahuana* (Ledig at al., 1997).

Table 7.1. The classification of rare British plants (from Rabinowitz et al., 1986).

Geographic distribution	*Wide*		*Narrow*	
Habitat specificity	Broad	Restricted	Broad	Restricted
Local population size				
Somewhere large	58	71	6	14
Everywhere small	2	6	0	3

This is an endangered species whose range retreated northward during the Holocene warming, which began 7,000 to 8,000 years ago. In the process the range fragmented and population sizes declined.

The contraction of ranges during long periods of climatic change is well known in species with good paleoecological records. During the past glacial epoch, species in the current vegetation zones of North America and Europe were forced southward into single or multiple refugia (Huntley and Birks, 1983; Delcourt and Delcourt, 1987). Species varied in their abundance within these refugia. Most of these species would be considered rare on the basis of distribution, if not on the basis of density. Trajectories to rarity taken by these species most likely were (2) and (3).

The spatial and temporal patterns of species decline are important to consider in conjunction with the overall issue of decline. A gradual change in climate may alter the relationships between species and the biota with which they interact, which in turn may lead to spatially heterogeneous patterns of decline. Pollinators and seed dispersers may change in their identity and abundance, and more importantly they may decline in their effectiveness. The failure of important mutualists may bring about a local or widespread downturn in species abundance (Bond, 1995). The breakdown of biotic interactions may be accelerated by range fragmentation and a reduction in local population sizes (Rathcke and Jules, 1993; Olesen and Jain, 1994). Competitors and herbivores also may change in their identity and abundance in ways that would contribute to local or regional downturns in species abundance (Geber and Dawson, 1993; Ayres, 1993)

The rise of a pathogen or pest to epidemic proportions is another form of environmental change. Two such epidemics were the chestnut blight (introduced from Asia to New York in 1904) and the Dutch elm disease (introduced from Europe into Ohio in 1930; Patterson and Backman, 1988). Their spread caused a progressive decline throughout the ranges of the American chestnut (*Castanea dentata*) and the American elm (*Ulmus americana*). The paleoecological record indicates that eastern hemlock (*Tsuga canadensis*) declined rather suddenly about 5,000 years ago and was absent or in low abundance for about 1,000 years (Davis, 1981, 1983). An insect pathogen may have been responsible for the decline (Bhiry and Filion, 1996). It took eastern hemlock 2,000 years to recover to its former abundance (Fuller, 1998). The aforementioned bottleneck may explain why the species has a low level of genetic diversity (Zabinski, 1992).

The rate of decline for a species depends on the magnitude of environmental change per unit time. Climatic changes are likely to be much slower than biotic changes, and thus will have a smaller impact on the abundance and range of a species per unit time than biotic changes. With regard to the latter, plants dependent on specialized mutualists are likely to undergo a more rapid rate of decline than species dependent on generalists. Species capable of self-fertilizing will suffer less than obligate outbreeders as pollinator service erodes.

The causes of species decline are complex and multifarious (Gaston, 1994). Without experimentation, drawing inferences about specific causal factors is inherently dangerous and may lead to ill-founded conclusions. Nevertheless, there are many assessments of why species have declined within the past century based only on observation. One such assessment is for Finnish plants as described by Lahti et al. (table 7.2).

The human element in species decline is clear. Thompson and Jones (1999) demonstrated this in an interesting way. They determined the relationship between the number of scarce plant species that have gone extinct since 1970 and the log of human population density in British vice-counties. There was a rather strong correlation ($r =$ ca. 0.60) between these variables. Eight of the 10 vice-counties that had lost the lowest number of scarce plants were in sparsely populated areas. Thompson and Jones argue that population density is an indicator of many aspects of land use that have a direct or indirect effect on species persistence. This is contrary to the popular belief that the main cause of the decline of British plants was agriculture.

Habitat and biotic changes are occurring with such speed that the rate of species decline within the past few centuries must be very much faster than occurred in the absence of humans. Thus we should not equate the dynamics of contemporary demise with that which occurred in prehuman times. When natural rates of decline are slower, species have much more opportunity to adapt to different habitats and to disperse to suitable sites. The potential for dispersal to favorable sites recently has been reduced because many corridors have been destroyed or fragmented. As a result climate tracking now may be limited or impossible, and many declining species are confined to their present geographic ranges and are more subject to extinction over short time spans.

Ultimately the pattern and rate of species decline in the face of a deteriorating environment depend on the complex interactions between the intrinsic features of species (e.g., environmental tolerances, resource requirements, and life history, demographic and dispersal properties). The attributes of their extrinsic environment that impact distribution and abundance also affect their pattern and rate of decline. Thus the patterns and rates of species decline are expected to be highly idiosyncratic, and indeed are, as seen from the fossil pollen record (Davis, 1989; Webb, 1987; Delcourt and Delcourt, 1987). Closely related species with similar ranges have the best chance of having similar de-

Table 7.2. Factors which ostensibly contribute to the rarity of eighty-three threatened Finnish vascular plants (from Lahti et al., 1991).

	Class		
	Vulnerable	Endangered	Extinct
Climate	32	25	7
Edaphic factors	27	17	1
Establishment	3	6	0
Dispersal	1	2	3
Hybridization	3	3	0
Population longevity	0	1	1

cline dynamics. Intrinsic characteristics of species that influence their ecological interactions and geographical distribution may be inherited from a common ancestor (Brown et al., 1996).

Genetic Aspects of Species Decline

The decline of species may have a major impact on their amount and organization of genetic variation. In general, the amount of variation is expected to fall as the effective size of a species declines. The level of differentiation among populations is expected to increase as ranges become more fractured, interpopulation distances increase, and populations become smaller (Holsinger, 1993; Young, Boyle, and Brown, 1996). If species decline is associated with an increase in local extinction and colonization, global genetic diversity is lost more rapidly than when populations are stable (McCauley, 1993). At the local population level, reduction in size is likely to be accompanied by an erosion of genetic variation due to increased genetic drift and elevated inbreeding.

Genetic Diversity

We would like to document genetic changes in different types of species as they decline. Typically this process is so slow that the task may be impossible. Studying species that are declining rapidly because of rampant, contemporary habitat destruction is not likely to be particularly informative in the short term. The effects of isolation and small population size may take several generations to be fully manifested.

Some understanding of the genetic attributes of declining species can be gained from studying species that contracted during the past several thousand years. The Chihuahua spruce is one such species. The species was once widespread in Mexico and has migrated northward in Mexico coincident with the Holocene warming 7,000 to 8,000 years ago. Isolated populations containing from 15 to 2,400 trees occur in the Mexican states of Chihuahua and Durango.

Ledig et al. (1997) surveyed the genetic property of its populations. Eleven loci of 24 allozyme loci are invariant. Genetic diversity (i.e., expected heterozygosity) ranges from 0.05 to 0.13 among populations with an average of 0.09. This value is within the range observed in the genus and in conifers in general. The percentage of polymorphic loci varies from 17 to 42%. About 75% of the variation is within populations and 25% among populations. The number of migrants per generation, Nm, is ca. 0.76. This is an order of magnitude lower than values reported for other conifers. The fixation index (F_{is}; the observed heterozygosity relative to that expected with random mating), is estimated to be 0.18. This is the highest value reported in conifers. It suggests that, on average, the parents of trees in the present generation were more closely related than half-siblings.

In summary, the Chihuahua spruce is not genetically depauperate, but is somewhat inbred. Forty-five percent of the seeds are empty, which ostensibly is the result of inbreeding.

Washingtonia filifera, the California fan palm, is another species that was once widespread but is now highly restricted. Formerly abundant along the California coast and what is now the Mojave Desert, it is restricted to isolated sites around the Colorado Desert of southern California. Based on an electrophoretic analysis of 16 loci, McClenaghan and Beauchamp (1986) found that the species has little genetic variation. The mean gene diversity across loci is 0.017. The average population was polymorphic at only 0.098 of its loci, and the average individual is heterozygous at 0.009 of its loci. There is little evidence of inbreeding. Surprisingly, populations are rather similar, with only 2.3% of the variation residing between them. This may reflect the relatively recent founding of some populations.

Some insights into the genetics of declining species also may be obtained from comparing species restricted to refugia with widespread congeners. This assumes that restricted species were once widespread and had genetical attributes similar to the present widespread species. The genus *Polygonella* affords the opportunity to compare widespread and restricted species. Lewis and Crawford (1995) surveyed populations of 11 species for allozymic diversity. Seven species are very narrowly distributed in the southeastern United States or in Texas; the remainder are broadly distributed in the southeastern United States. Contrary to expectations, the more restricted the range, the more genetic diversity is maintained. When pairs of closely related species are compared, the species with the more southerly distribution have higher levels of gene diversity and proportion of polymorphic loci. Lewis and Crawford propose that this pattern arises from the long-term geographical stability of the southern populations. The species that migrated to the north as they tracked environmental change presumably lost genetic variation over the past several thousand years.

Polygonella is not unique in having higher levels of variation in populations close to refugial areas. Higher levels of variation are found in southern populations in the wetland herb *Helonias bullata* (Godt, Hamrick, and Bratton, 1995), *Pseudotsuga menziesii* (Li and Adams, 1989), *Abies grandis, Abies concolor* (Steinhoff, Joyce, and Fins, 1983), *Pinus monticola* (Critchfield, 1984), and *Thuja plicata* (Copes, 1981).

There have been several comparisons of geographically restricted species and their widespread congeners. As summarized by Karron (1991), the former tend to have less genetic variation than the latter. However, we cannot assume that the restricted taxa were ever widespread and that they lost variation as they declined. Indeed in genera such as *Clarkia, Gaura,* and *Layia* some restricted taxa ostensibly have evolved recently.

Occasionally a widespread species will have little genetic variation, whereas other widespread congeners will have much more. This is the case in *Pinus* where *P. resinosa* has only 11% polymorphic allozyme loci and averages 2.0 alleles per polymorphic loci (Allendorf, Knudsen, and Blake, 1982) versus 92% polymorphic loci and 3.2 allele per polymorphic loci in *P. contorta* (Yeh and Layton, 1979) and 76% polymorphic loci and 3.1 alleles per polymorphic loci in *P. rigida* (Guries and Ledig, 1982). The unusually low level of variation in *P. resinosa* may be the result of a bottleneck in its recent history.

Genetic Variation Versus Population Size

Species decline eventually leads to entities that are rare and that are made of small isolated populations. Considerable attention has been given to the genetic properties of such species and the impact of proximal factors acting on local populations. Consider first the relationship between population size and genetic variation. Based on population genetic theory, small isolated populations will be subject to genetic drift and inbreeding; the smaller the population the greater the effects are likely to be. Through the loss of favorable alleles and as the result of inbreeding depression, small populations may lose some adaptability and be more prone to extinction than larger ones.

The amount of allozyme variation was determined in populations of *Salvia pratensis* ranging from 5 to 1,500 plants (Ouborg and van Treueren, 1994), populations of *Scabiosa columbaria* ranging from 14 to 100,000 plants (van Treuren et al., 1991), *Gentiana pneumonanthe* ranging from 1 to 50,000 plants (Raijmann et al., 1994), and populations of *Eucalyptus albens* ranging from 14 to more than 10,000 plants (Prober and Brown, 1994). In all species there is a significant positive correlation between the logarithm of population size and both the proportion of polymorphic loci and the number of alleles at those loci (figure 7.2). In Chihuahua spruce there is a very strong correlation ($r = 0.93$) between genetic diversity and the logarithm of population size (Ledig et al., 1997). The correlation between the percentage polymorphic loci and population size was not significant ($r = 0.61$). In *Washingtonia filifera* the percentage of polymorphic loci is significantly correlated ($r = 0.72$) with population size. Significant positive correlations between various measures of allozyme diversity and population size also occur in *Geum radiatum* (Hamrick and Godt, 1995) and *Gentianella germanica* (Fischer and Matthies, 1997).

Fischer and Matthies (1998b) extended their studies of genetic variation in *Gentianella germanica* (Gentianaceae) to include RAPDs (random amplified polymorphic loci) as well as allozymes. They identified 54 RAPD profiles among plants from 11 populations whose sizes ranged from 40 to 5,000 reproductive individuals. There is a high rank correlation ($r = 0.78$) between the amount of molecular variation within populations and the log of population size (figure 7.3). This relationship is fostered by the large distances between populations and by the low level of gene flow between populations. The median distance to the closest population sampled in this study was ca. 3 km. The higher the molecular variance of populations the greater the mean seed-set per plant (figure 7.3).

The amount of phenotypic variation in relation to population size was estimated from common garden trials in *Scabiosa columbaria, Salvia pratensis* (Ouborg, van Treuren, and van Damme, 1991) and *Gentiana pneumonanthe* (Oostermeijer et al., 1994). The amount of phenotypic variation is positively correlated with population size in the first two species, but statistically significant only in *S. pratensis*. In *Gentiana pneumonanthe* there is no relationship, but the trend is toward higher variation in small populations.

A positive correlation between genetic diversity and population size is not the rule. No relationship between these variables was found in *Eucalyptus caesia*

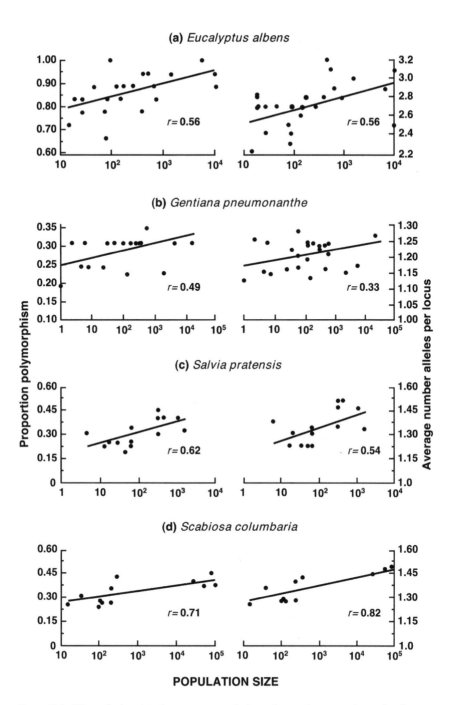

Figure 7.2. The relationship between population size and proportion of polymorphic loci and average number of alleles per locus in four species (from a, Prober and Brown, 1994; b, Raijmann et al., 1994; c, van Treueren et al., 1991. All redrawn with permission of Blackwell Science Ltd.).

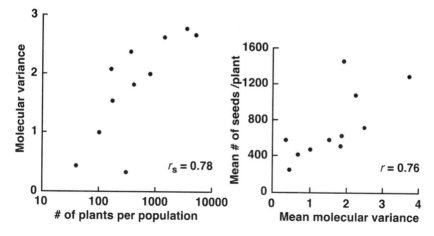

Figure 7.3. The relationships between number of plants per population and molecular variance (left) and the mean molecular variance and the mean number of seeds per plant (right) in *Gentianella germanica* (from Fischer and Matthies, 1998. Redrawn with permission of the Botanical Society of America).

(Moran and Hopper, 1983), *Eucalyptus pendens* (Moran and Hooper, 1987), *Acacia anomala* (Coates, 1988), *Stylidium coroniforme* (Coates, 1992), and *Centaurea corymbosa* (Colas, Olivieri, and Riba, 1997).

The diversity of allozymes and quantitative traits is not likely to have a substantial bearing on the immediate fitness of a population. However, the same is not the case for allele diversity at the self-incompatibility (S) locus. The fewer the number of alleles, the greater the restriction on the number of pairs of plants that can effectively interbreed, especially if the species has sporophytic incompatibility. Byers and Meagher (1992) demonstrated in a computer simulation that populations with 25 or fewer individuals are unable to retain large numbers of S-alleles, if they have this type of incompatibility.

We see evidence of few S-alleles in small populations of *Hymenoxys acaulis* var. *glabra*; crosses within populations are unsuccessful, while crosses between populations are successful (De Mauro, 1991). Reduced seed-set in intrapopulation crosses in *Aster furcatus* also suggests low S-allele diversity (Reinhartz and Les, 1994). This species has limited variation at allozyme loci (Les, Reinhartz, and Esselman, 1991). Six to 8 S-alleles occur in small isolated populations of *Carthamus flavescens* (Imbrie and Knowles, 1971; Imbrie, Kirkman, and Ross, 1972). All of the aforementioned species have sporophytic incompatibility.

The relationship between S-allele diversity and population size remains to be determined. The effect of genetic drift in small populations is counterbalanced in part by frequency-dependent selection at the S-locus, where the rarer an allele becomes, the greater is its selective advantage (Wright, 1939). Thus, we would expect S-gene diversity to be less associated with population size than neutral gene diversity.

The lack of a connection between S-gene and allozyme diversity is seen at the species level in the narrow endemic *Oenothera organensis*, which is a native of the Organ Mountains in southeastern New Mexico. The species may comprise fewer than 5,000 individuals divided into numerous small populations. Emerson (1939, 1940) sampled four populations and identified more than 45 self-incompatibility alleles in this species. The spatial distribution of these alleles is rather homogeneous. Levin et al. (1979) investigated allozyme diversity at 15 loci from 10 well-spaced populations. Only one locus (*Mdh*) is polymorphic. The level of random outcrossing made from the frequencies of homozygotes and heterozygotes is close to 1.00 in most populations, indicating that local populations are not inbred. Gene exchange between populations ostensibly mitigates inbreeding. The estimate of F_{st} is 0.14; thus gene frequency differentiation between populations is minor. Pollen is carried within and among populations by strong-flying hawkmoths. Seeds have no special adaptations for dispersal, but long-distance transport probably is accomplished by deer that heavily browse inflorescences and capsules.

The problem with relating genetic variation to population size is that the ages and demographic histories of populations typically are unknown. If populations have been small for a short time, it may be too soon for the effects of genetic drift to be manifested. This is illustrated in the Mauna Kea silversword *Argyroxiphium sandwicense* (Friar, Robichaux, and Mount, 1996). The species is a monocarpic perennial that does not flower until it is 30 to 50 years old. It was abundant on the Mauna Kea volcano on the island of Hawaii before the introduction of ungulates, which occurred in the late 18th century. The silversword population gradually declined to its recent census of 46 plants. The proportion of polymorphic loci and level of heterozygosity is similar to that of a large conspecific population growing on the Haleakala volcano on the island of Maui. The size of the latter has exceeded 4,000 plants since records have been kept and recently surpassed 60,000 plants.

Inbreeding Depression

If populations have been small for a long time, selection is expected to reduce the genetic load and thereby inbreeding depression in them (Barrett and Charlesworth, 1991). Deleterious alleles would be purged, eventually leading to a net increase in average fitness. However, when inbreeding depression is the result of deleterious recessive alleles with small effects on fitness, the purging of these alleles may be hampered by genetic drift (Charlesworth and Charlesworth, 1987) and high mutation rates to deleterious alleles (Johnston and Schoen, 1995), but promoted by self-fertilization (Charlesworth, Morgan, and Charlesworth, 1990). If populations only recently have become small, their response to inbreeding would be similar to large populations in the same area. In these populations, the effects of inbreeding and genetic drift will accumulate slowly. The resulting increase in homozygosity is expected to proceed slower and later for fitness loci than for neutral loci (Barrett and Kohn, 1991).

Consistent with theory, there is a positive association between the level of random outcrossing and inbreeding depression upon forced selfing in some species with mixed mating systems such as *Clarkia tembloriensis* (Holtsford and Ellstrand, 1990), *Eichhornia paniculata* (Barrett and Kohn, 1991), and *Collinsia heterophylla* (Mayer, Charlesworth, and Meyers, 1996). However, other studies report weak or no associations between mating system and level of inbreeding depression (e.g., *Kalmia latifolia*, Rathcke and Real, 1993; *Decodon verticillatus*, Eckert and Barrett, 1994). Perhaps the lack of association is not surprising given that several other factors besides the mating system also determine the magnitude of inbreeding depression (Charlesworth and Charlesworth, 1990; Charlesworth, Morgan, and Charlesworth, 1990, 1991). These include the kind and degree of epistasis among fitness-determining loci, the strength of selection, and the mutation rate to deleterious alleles.

We also would expect populations long differing in size to display different levels of inbreeding depression because small populations should be subject to the greatest level of inbreeding (Barrett and Kohn, 1991; Ellstrand and Elam, 1993). However, in populations of *Salvia pratensis* (Young et al., 1996), *Scabiosa columbaria* (van Treuren et al., 1993), and *Senecio integrifolius* (Widen, 1993) there is no relationship between population size and inbreeding depression. Ostensibly, the smaller populations reached their size recently, and there was insufficient time for the purging of deleterious alleles. A few generations of selfing may have little effect on the magnitude of inbreeding depression as shown in the predominantly outcrossing *Mimulus guttatus* (Latta and Ritland, 1994; Dudash, Carr, and Fenster, 1997).

The size of a population may be measured in terms of the number of individuals it contains or in its effective size (N_e). The effective size of populations refers to the size of an idealized population that would have the same genetic properties relative to genetic drift and inbreeding as that observed for a real population (Wright, 1931, 1938). If we draw two genes from a population whose effective size is less than the actual size, we are more likely to get two copies of the same gene than we would expect when the chance of drawing two copies of the same gene from the standing crop of N plants is $(1/2N)^2$.

The effective sizes of the populations in the aforementioned examples are unknown. The relationship between genetic variation or inbreeding depression and population size probably would have been stronger if the effective size were used.

Newman and Pilson (1997) conducted an intriguing study on the relationship between inbreeding depression and effective population size in synthetic populations of *Clarkia pulchella*. They generated 16 high N_e and 16 low N_e populations that differed in effective size for three years. The high populations were founded with one individual from each of 12 maternal families. The low populations were founded by four full siblings from each of three families. Although the difference between the two groups of populations is small the first two years, in the third year 98% of the flowering plants survive to fruit in the high N_e treatment versus 50% in the low N_e treatment. Moreover the fitness (survival to flowering × mean fruit number per plant) of the high N_e populations is significantly greater than that of the low populations.

The magnitude of inbreeding depression in small populations also may be a function of their ploidal level. Lande and Schemske (1985) predicted that auto- and allotetraploidy should substantially reduce inbreeding depression caused by recessive lethal and sublethal alleles in outcrossing populations. The equilibrium inbreeding depression of a tetraploid should be nearly half that of the diploid progenitor. For alleles that are partially recessive, they predicted that diploid and tetraploids would not differ in inbreeding depression.

The most direct test of Lande and Schemske's prediction in wild plants was made by Husband and Schemske (1997). They estimated inbreeding depression in the greenhouse for two diploid and three tetraploid populations of *Epilobium angustifolium* (Onagraceae) from seed-set, germination, survival, and dry mass at nine weeks. The two ploidal levels have similar rates of selfing (close to 0.45), so differences between them cannot be attributed to divergent mating systems. The tetraploid populations display less inbreeding depression than the diploids. The cumulative inbreeding depression calculated from these stages average 0.95 in the diploids versus 0.67 in the tetraploids.

The relationship between ploidal level and inbreeding depression has been sought in several crop species. Synthetic tetraploids of clover (Townsend and Remmenga, 1968), maize (Alexander, 1960), and crested wheatgrass (Dewey, 1969) have less inbreeding depression than their respective diploid counterparts. Conversely, polyploids have higher levels of depression in crested wheatgrass (Dewey, 1966) and orchardgrass (Kalton, Smit, and Leffel, 1952). Regardless of the relationship, there is a clear effect of ploidy on inbreeding depression in all of the species.

Given the data presented in this section, what are the genetic correlates of species decline? Decline without an approach to extinction seems to have only a small effect on genetic variation within a species as judged from allozyme data, which are presumed to reflect neutral polymorphisms. However, decline may affect variation for traits immediately related to fitness, because there is little evidence in outcrossing species that the level of variation in allozymes within or among populations is indicative of the level and organization of variation for other attributes (Hamrick, 1989).

Interpopulation Hybridization

If small populations are subject to pollen flow from their neighbors and if pollen flow is important in seed production, then the quality of interpopulation hybrids will determine the long-term impact of pollen import. Crosses between populations may yield hybrids that are superior, inferior, or equivalent to plants from intrapopulation crosses. If hybrids are superior the prospects for population persistence are enhanced, whereas if hybrids are inferior these prospects may be diminished.

Interpopulation crosses in some rare species yield hybrids that are superior to both parents. Van Treuren et al. (1993) compared the germination, survival to flowering, and number of flower heads of between population progeny (BPP) of *Scabiosa columbaria* (Dipsacaceae) with those of within population progeny (WPP) in the greenhouse. Six populations were intercrossed in the investigation. On average, germination in the BPP is 89% versus 78% in the WPP, and seedling to adult survivorship is 33% in the BPP vs. 26% in the WPP. The aboveground biomass averages 3.0 grams for BPP vs. 2.4 grams for WPP under competitive conditions. The differential is somewhat less in noncompetitive trials. The mean number of heads per plant averages 39.3 in the BPP and 32.7 in the WPP. Interpopulation hybrid advantage also has been demonstrated in *Ipomopsis aggregata* (Heschel and Paige, 1995), *Sabatia angularis* (Dudash, 1990), *Salvia pratensis* (Ouborg and van Treuren, 1994), *Gentiana pneumonanthe* (Oostermeijer, Altenburg, and Den Nijs, 1995), *Phlox drummondii* (Levin and Bulinska-Radomska, 1988), and *Eupatorium perfoliatum* (Byers, 1998).

Hybrid advantage may be affected by the environment. It often is enhanced under stressful conditions such as high temperature or drought (cf. Lesica and Allendorf, 1992). Thus it is likely that the relative fitness of the interpopulation hybrids in the aforementioned species would be greater under stressful conditions than benign ones.

Interpopulation hybrids are not always superior to local plants. For example, in *Gentianella germanica* the germination rates, rosette size of progeny, and survival to 175 days are higher for the progeny of 10 m crosses than those of progeny from interpopulation crosses (Fischer and Matthies, 1997). This outbreeding depression may even occur between distant plants within populations (Waser and Price, 1994). Outbreeding depression may be the result of local adaptation, in which different populations are adapted to different habitats, or intrinsic coadaptation, in which genes of one population are adapted to a different genetic environment than genes of another population (Templeton, 1986). In either case hybridization will disrupt functional genetic architectures.

Fenster and Dudash (1995) suggest that there may a relationship between the breeding system of species and the magnitude of outbreeding depression for a given crossing distance. Predominantly selfing species are expected to have greater outcrossing depression than outcrossers because the former may diverge in epistatic gene complexes more rapidly. The relationship between outcrossing depression and the breeding system remains to be determined.

The fitness of progeny from interpopulation crosses may vary with the distance between the populations. Neighboring populations are likely to be less divergent than distant populations, and thus crossing between neighboring populations is less likely to yield heterotic or depressed hybrids than will crossing between distant populations. For example, in *Streptanthus glandulosus* hybrids from nearby crosses have higher fertilities than hybrids from distant crosses (Kruckeberg, 1957). None of the crosses in the aforementioned studies was between neighboring populations, so we do not know the fitness of a normal spectrum of hybrids.

Moll et al. (1965) demonstrated that the fitness of progeny from interpopulation crosses also varies with the level of genetic divergence between maize strains. The maximum level of heterosis is in crosses between moderately divergent strains. Heterosis in hybrids between the most divergent strains is similar to that in the hybrids of the least divergent strains.

The fitness of progeny from interpopulation crosses also may be dependent on the level of inbreeding within populations, especially in outcrossing species. This is best illustrated using cultivars in conjunction with natural populations. In 1985 I compared the fitnesses of *phlox drummondii* cultivars ('White Beauty', 'Isabellina', 'Crimson Beauty', and 'Brilliant Rose') and hybrids in natural *P. drummondii* sites. Hybrid seed and cultivar seed were synthesized in greenhouse crosses. Seeds of each cultivar and F_1 hybrids between 'White Beauty' and 'Isabellina' and between 'Crimson Beauty' and 'Brilliant Rose' were planted in replicated blocks at three Texas sites. Germination, survival to flowering, the mean number of seeds per flowering plant, and the net reproductive rate were determined for cultivars and hybrids at each site.

Hybrids are superior to parents in all aspects of their life histories (table 7.3). The mean germination rate for hybrids is 0.60 versus 0.33 for the parental cultivars, and the mean survival in hybrids is 0.20 versus 0.13 in the cultivars. The mean number of seeds per reproductive plant is 12.0 in the hybrids compared to only 4.1 in the cultivars. On average the net reproductive rates of hybrids is 0.50 versus 0.25 for the cultivars. It is noteworthy that the hybrid with the highest net reproductive rate ('Crimson Beauty' × 'Brilliant Rose') is derived from cultivars whose mean performance is superior to the cultivars from which the other hybrid ('White Beauty' × 'Isabellina') is derived.

The 100% superiority of intercultivar *Phlox* hybrids relative to parental strains stands in stark contrast to the 20% superiority of interpopulation hybrids relative to natives (Levin and Bulinska-Radomska, 1988). If these data are representative, we would expect local populations which are most inbred to express the greatest relative level of heterosis in their hybrids. This relationship indeed is found in the Australian herb *Isotoma petraea* (Beltran and James, 1974) and in many crops (Frankel, 1983).

Table 7.3. Demographic estimates for *Phlox* cultivars and hybrids.

Site	Cultivar or Hybrid					
	WB	IS	WB × IS	CB	BR	CB × BR
(a) Emergence						
GONZALES	.55	.06	.47	.21	.56	.65
LYTLE	.30	.10	.57	.35	.45	.61
SARITA	.42	.06	.62	.24	.57	.69
Population Means	.42	.08	.58	.27	.49	.66
(b) Survival to reproduction						
GONZALES	.20	.16	.24	.47	.18	.28
LYTLE	.17	.13	.20	.10	.11	.40
SARITA	.00	.00	.02	.00	.00	.10
Population Means	.12	.10	.15	.19	.10	.26
(c) Mean numbers of seed produced by reproductive plants						
GONZALES	5.4	1.0	4.4	6.2	4.9	7.3
LYTLE	7.8	6.7	15.6	13.6	4.1	39.5
SARITA	.00	.00	2.67	.00	.00	2.9
Population Means	4.4	2.6	7.2	6.6	2.6	12.4
(d) Net reproductive rates						
GONZALES	.61	.10	.49	.61	.49	1.32
LYTLE	.39	.09	1.77	.48	.20	9.63
SARITA	.00	.00	.03	.00	.00	.20
Population Means	.33	.06	.76	.36	.23	3.72

WB = White Beauty, IS = Isabellina Yellow, CB = Crimson Beauty, BR = Brilliant Rose

Studies of interpopulation hybrid fitness typically have considered only the first generation. One study going beyond F_1 hybrids is that on maize referred to above (Moll et al., 1965). Heterosis in the F_2 generation is roughly half that observed in the F_1s at each level of divergence. A similar result was reported in *Isotoma petraea* (Beltran and James, 1974). This is what is expected in the absence of epistasis, because the F_2 generation has half the heterozygosity of the F_1 generation (Falconer, 1981).

Demographic Aspects of Species Decline

The ecological consequences of species decline, like the genetic consequences, depend on the original distribution and abundance of a species. Species with widespread and continuous distributions will be affected the most. Those with small, widely spaced populations will be affected the least. Decline in species with widespread and continuous distributions is likely to be manifested in smaller population size, fewer populations, and lower rates of migration between populations, at least in parts of the range where habitats are becoming less hospitable.

Seed Production

Outcrossing plants in small populations often produce fewer seeds per flower than plants in large populations. Reduced reproductive success within populations stems from insufficient pollination. Pollinators often form search images of the most favorable species, which is determined principally by the quantity and quality of the floral reward offered by a population (Levin, 1978b). When a plant species is locally rare relative to other possible host species, it is likely to be discriminated against or its pollinators will have a lower than normal level of floral constancy (the tendency to forage among conspecific plants, Rathcke, 1983; Aizen and Feinsinger, 1994a; Kunin, 1997). In either case, conspecific pollination suffers.

There have been few direct observations of pollinator service in relation to number of flowering plants. In *Nepeta cataria* the number of visits per flower per unit time by *Apis* and *Bombus* increases with population size; however, halictid bees respond to size in the opposite way (Sih and Baltus, 1987). Population size explains 74 to 85% of the visitation rate. Jennersten (1988) reported that in the Swedish plant *Dianthus deltoides*, small isolated populations receive less visitation than large populations. Conversely, patch size in Swedish populations of *Viscaria vulgaris* has no overall effect on visitation rate (Jennersten and Nilsson, 1993). According to Rathcke (1983), each species has an optimal patch size regarding pollinator visitation.

Seed-set per flower often is a positive function of population size. As population size declines, the decline in seed-set may be disproportionate. When populations fall below some threshold size, reproduction may be lacking. This is referred to as the Allee effect (Allee, 1949; Lande, 1988). Relatively low levels of seed-set per flower occur in small populations of numerous species (Levin,

1995). Seed-set per flower in these populations usually is 25 to 50% less than for plants within large populations. There may be a substantial correlation between reproductive performance and population size. For example, in *Gentianella germanica* the correlation between seeds per plant and the log of population size is $r = 0.54$ (Fischer and Matthies, 1998a). In *Lythrum salicaria* the correlation between these variables is $r = 0.75$ (Ågren, 1996). Plant size is independent of population size in both species. In *Senecio integrifolius* the correlation between the percentage of ovules setting seed and the log of the number of flowering stems is $r = 0.87$ (Widen, 1993). A positive correlation between fecundity and population size also has been reported for wind-pollinated species (*Fagus sylvatica*; Nilsson and Wästljung, 1987).

When population size is very small no seeds may be produced. For example, in the rare Australian *Banksia goodii*, populations less than $25m^2$ did not produce any seeds (figure 7.4; Lamont, Klinkhamer, and Witkowski, 1993). In the rare *Senecio integrifolius*, one population of $N = 3$ did not set seed, whereas populations over 1,000 plants had seed-sets exceeding 70% (Widen, 1993).

Aizen and Feinsinger (1994a) studied pollination levels and fruit and seed production in 16 species in a dry subtropical forest in northwestern Argentina that contained small and large fragments and continuous tracts. The pollination systems and growth forms of the species are quite diverse. All are outbreeders. The median pollination level (measured in terms of pollen grains deposited on stigmas or pollen tubes per pistil) and seed output are less in fragments than in forests. The responses to population size vary among species. Half of the species show a positive relationship between the number of pollen tubes at the base of the style and the population size. In nearly half the species fruit-set varies positively with population size; and in about 35% of the species seed-set varies positively with population size.

The effect of population size on reproductive success may be attributed in part to a change in the pollinator fauna. Aisen and Feinsinger (1994b) assessed

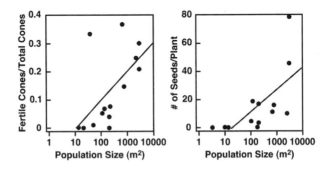

Figure 7.4. The relationship between population size and (left) proportion of cones that are fertile and (right) number of seeds per plant in *Banksia goodii* (from Lamont et al., 1993. Redrawn with permisson of Springer-Verlag).

the responses of the pollinator assemblage to habitat fragmentation for two of the species considered in the aforementioned study, *Prosopis nigra* and *Cercidium australe*. The frequency and taxon richness of native-flower visitors to both species declines with decreasing forest-fragment size. Likewise, a fragmentation-related decline in taxonomic diversity of pollinators has been documented in *Dianthus deltoides* (Jennersten, 1988). Powell and Powell (1987) found fragmentation-related declines in diversity and density of euglossine bees attracted to scent traps in 100-, 10-, and 1-hectare forest remnants near Manaus, Brazil. Sowig (1989) reported that in *Aconitum napellus* and *Symphytum officinale* short-tongued species of bumblebees are most dominant in large patches, whereas in small patches middle- and long-tongued bumblebees prevail.

If plants are facultatively autogamous, small populations may have a different breeding behavior from large conspecific populations. Lower pollinator visitation to small populations often results in higher rates of self-fertilization (e.g., *Carduus nutans*, Smyth and Hamrick, 1984; *Nepeta cataria*, Sih and Baltus, 1987; *Cavanillesia plantanifolia*, Murawski et al., 1990). The higher selfing rates do not indicate, however, that an increase in self-seed production fully compensates for the loss of cross-seed production. An increase in selfing in *Dianthus deltoides* (Jennersten, 1988) and *Lupinus nanus* (Karoly, 1992) has been accompanied by net loss of seed production.

In small populations subject to pollinator limitation, self-fertile morphs in species with heteromorphic incompatibility have an unusual advantage. Consider the situation in a remnant population of *Primula sieboldii*, which has numerous pins and thrums in a 1 : 1 ratio and one short homostyled plant (Washitani et al., 1994). The percentage of flowers setting fruit in the pins, thrums, and homostyle morphs are 20, 9, and 82, respectively. The mean seed-sets per flowers are 7.6. 2.1, and 32.0, respectively. These differentials could lead first to the loss of the thrum morph and eventually to the replacement of heterostylous, self-incompatible plants by homostylous, self-compatible ones (Washitani, 1996).

The reproductive success of self-incompatible species in small populations may be compromised by a paucity of self-incompatibility alleles. When this is the case, the number of potential mates is far fewer than the number of flowering plants, especially if the species has sporophytic incompatibility. A paucity of S-alleles ostensibly explains the low mating success among plants in small populations of *Hymenoxys acaulis* var. *glabra* (De Mauro, 1991) and *Aster furcatus* (Reinhartz and Les, 1994).

Finally, chance deviation from a 1 : 1 ratio of females to males in dioecious species or of pins and thrums in distylic species may reduce reproductive success in small populations. The smaller the population, the greater the deviation, from a 1 : 1 ratio is likely to be; and the greater the deviation the greater the percentage of crosses will be within a morph and thus not productive. The heterostylous *Amsinckia grandiflora* (Boraginaceae) occurs only at one site in central California. The population is small and the ratio of pins to thrums fluctuates substantially from year to year. When the ratio is very skewed, most pollination is ineffective and seed-set may be very low (Huenneke, 1991).

A distinction needs to be made between population size and plant density. Small populations may have plants spaced closely or far apart as may large populations. Kunin (1997) used some creative planting designs to separate the effect of size and density in garden populations of *Brassica kaber*. He planted fan-shaped density arrays at spacings ranging from 10.5 cm to 10 m apart. Plants at low density have markedly lower rates of pollinator visitation and lower seed-set than plants at moderate to high densities. In another experiment where density was held constant, Kunin planted triangular populations ranging in size from 3 to 78 plants. Surprisingly, neither visitation rates nor seed-set are affected by population size. As summarized by Kunin (1997), plant reproductive success typically is a positive function of plant density, whereas reproductive success may or may not vary with population size. Clearly, small populations of well-spaced plants are especially vulnerable to reproductive failure.

The reproductive success of small populations depends in part on their proximity to other populations. If other populations were in close proximity and pollen immigration was substantial, any liability inherent in small size might be overcome. Groom (1998) utilized roadside populations of *Clarkia concinna* as an experimental system. She determined the proportion of stigmas receiving pollen and the seed-set in very small populations (less than 10 plants), small populations (10–50 plants), and large populations (more than 50 plants), while noting the proximity of the nearest population. When very small populations were more than 20 m from a neighbor, almost no stigmas receive pollen and almost no seeds were set. For small populations the distance threshold beyond which reproductive success is meager is ca. 100 m. The reproductive performance of large populations is independent of distance to neighboring populations.

Progeny in small populations may have a higher incidence of genetic defects than progeny in large populations, because of the more frequent coupling of deleterious, recessive alleles. Population-size dependent rates of phenotypic abnormality (e.g., chlorophyll deficiency) have been described in *Pinus radiata* (Squillace and Krause, 1963; Bannister, 1965).

Plant Performance

Whereas we know much about per capita reproduction in relation to population size, we know little about germination and survival in relation to population size. It is likely that plants in small populations would be inferior to those in large populations in these respects, all else being equal, because the former would be more inbred. Heschel and Paige (1995) showed that in the self-incompatible *Ipomopsis aggregata* seed size and germination and plant regrowth following simulated herbivory are less in plants from small populations than those in large populations. Plants from the small populations also have higher mortality following simulated herbivory. The mean seed-set in small populations increases from ca. 25% to ca. 50% when plants were hand-pollinated with pollen from distant sources. Hand-pollination with pollen within populations yields seed germination percentages like those obtained with natural pollination. Distant pollination of a large population has no impact on germination. Seed mass also increases with distant pollination in small populations. Ostensibly inbreeding depression is responsible for the reduced performance in small populations.

Menges (1991) studied germination percentages in small (less than 150 plants) and large populations (more than 150 plants) of the rare hummingbird-pollinated *Silene regia* (Caryophyllaceae). Seeds collected from plants in large populations have the highest germination percentages. Perhaps plants from smaller populations received few pollinators visits per flower, which would lead to higher levels of self-fertilization than in plants of large populations.

Silene regia also has been the subject of a viability analysis. Menges and Dolan (1998) collected demographic data from 16 populations for up to seven years and used matrix projection methods to describe population growth rates. There is a positive correlation ($r = 0.48$) between population size and the finite rate of increase. The univariate relationships between the species' genetic variation and demographic viability are weak. Genetic variation has positive effects on finite rates of increase and negative effects on extinction probability in multivariate analyses that included the effects of management.

Edge Effects

During the decline of species, large continuous populations often are fragmented into several smaller ones. Little attention has been given to properties of these fragments or local populations other than their size. In addition to being smaller, the fragments differ from the larger system in the ratio of population perimeter to interior area. This ratio will be much higher in the fragments, especially if they are long and narrow. A high ratio makes a population more vulnerable to gene flow from extraneous populations because border plants usually receive less local pollen than interior plants, especially in wind-pollinated plants (Pedersen, Johansen, and Jorgensen, 1961; Knowles and Baenziger, 1962). Therefore plants may have lower reproductive success than interior plants. Consequently, plants near the periphery of populations are likely to produce higher levels of interpopulation hybrids as demonstrated in corn (Jones and Brooks, 1950), smooth bromegrass (Knowles and Ghosh, 1968), cotton (Green and Jones, 1953), and alfalfa (Pedersen et al., 1969).

The pattern of pollen receipt has important implications for the spread of introduced genes. Using computer simulation, Levin and Kerster (1975) showed that if an introduced gene is favored, and if pollen and seeds are dispersed narrowly within populations, edge receipt of alien pollen results in a much slower rate of gene substitution than if alien pollen receipt was random within populations. If the introduced gene is detrimental, edge receipt will lead to a lower equilibrium frequency than if pollen receipt is random within the population (Levin and Kerster, 1975).

Although border plants may receive less pollen per stigma than interior plants, they are not spatially isolated in the sense that peripheral populations are isolated from interior populations. They cross not only with other border plants, but also with those in the interior as is illustrated in Handel's (1982) experiments on melons.

Additional evidence of crossing between edge and border plants comes from the foraging behavior of honeybees in populations of *Lythrum salicaria* (Levin, unpubl.). In the interior of dense populations, honeybees usually fly from a plant to one of its near neighbors. Honeybees on or near the periphery often fly well beyond the population, as if looking for more plants, and then loop back to peripheral plants or plants in the interior. As a result, the average interplant distance of flights from peripheral plants is about ca. 150% greater than flights from interior plants.

In insect-pollinated plants the pattern of mating within small populations also may depend on the ratio of the periphery to the interior area. Bateman (1947) demonstrated that gene flow distances from a core of radishes to plants arranged in arms radiating from the core were an inverse function of the number of plant rows in the arm. In a related experiment, Manasse (1992) found that gene dispersal distances in mustard were greater in one-dimensional plant arrays (i.e., a single file of plants) than in two-dimensional arrays. Of particular importance is the observation that honeybees foraging on one-dimensional mustard arrays have biased directional movement rather than foraging at random (Morris, 1993). If pollen from one plant is retained on a honeybee as it visits several different plants, as is typically the case (Levin and Kerster, 1974), directional flights greatly expand the axial distance between crossbreeding plants and thus the gene dispersal distance.

Extinction and Its Approach

Extinction has been considered at two levels: (1) the extinction of local populations, and (2) the extinction or near extinction of species. Of particular interest are those studies that provide insights into genetic and/or demographic factors that bring about the local or general demise of a species.

Local Extinction

Very little is known about the local extinction of populations across a broad time span, especially in relation to population attributes. One notable study is that by Ouborg (1993), who compared species composition and abundance at 143 sites along the Dutch Rhine system for 1956 and 1988. The number of extinction and colonization events was determined for 15 species at each site. This information was considered in light of the nearest site that contained conspecific plants and the number of sites within 1.5 km that contained conspecific plants.

The number of extinctions between 1956 and 1988 was much greater than the number of colonizations for each species. Extinct populations of most species were on average farther away from their nearest neighbors and were smaller in 1956 than their extant counterparts. Some species such as *Medicago falcata* and *Eryngium campestre* reacted to both population size and isolation. Others reacted to population size, but not the proximity of neighboring populations (e.g., *Cynodon dactylon*). The opposite was true for *Plantago media*. The population extinction of *Scabiosa columbaria* was unrelated to population size or neighbor distance.

Groom (1998) considered extinction in relation to population size and isolation in *Clarkia concinna*. Roadside populations were monitored in central California between 1992 and 1997. Populations were scored as very small (1–10 plants), small (11–50 plants), and large (more than 50 plants). Three categories of isolation were recognized (less than 26m, 26–100m, more than 100m). There were 28 extinctions among 211 populations during the six-year study. On average the rate of extinction is highest in very small populations (ca. 50%) and lowest in large populations (ca. 5%). Taking into account population size and isolation, the highest extinction level (75%) is in very small populations isolated by at least 100m. Catastrophic disturbance (erosion and human influence) is the leading cause of extinctions in all but the very small populations, where low reproductive success is pivotal.

Another study of the role of population size on extinction involved the genesis of populations that differed in effective size. Newman and Pilson (1997) established 16 high and 16 low effective size populations of the annual *Clarkia pulchella* in a prairie habitat typical for the species. All high and low populations survived the first two generations (years) of the experiment. However, 15 of the high, but only 11 of the low populations survived the third generation; and 12 of the high, but only 5 of the low survived the fourth generation. Thus, after three generations 75% of the high N_e populations were still extant versus only 31% of the low N_e populations. Since the numbers of plants in the initial populations were the same, differences in the survivorship of the two groups of populations must be a function of inbreeding and genetic drift.

Species Extinction

At some point in the decline of species their existence may become tentative. They may totter on the brink of extinction. These species are characterized by small numbers of populations. The number of known populations for 91 species listed as endangered or threatened by the U.S. Fish and Wildlife Service (figure 7.5) was compiled by Schemske et al. (1994). The majority of species have five or fewer populations.

In order to understand the possible fate of endangered species, ecological and genetic studies should be brought to bear. I will consider some species at the brink and what both approaches foretell of their future.

Centaurea corymbosa is endemic to the Massif de la Clape near Narbonne in southern France, where it occupies an area less than 3 km². The species is known from six populations. In June 1995, Colas et al. (1997) found 494 reproductive plants and estimated the total population size to be roughly 6,500 individuals. They analyzed enzyme polymorphisms at the six known populations. Of 19 loci tested, only 5 are polymorphic; and they are diallelic. Genetic diversity as measured by Nei's index is a very low 0.074. There is no relationship between population size and the genetic diversity of local populations. The overall F_{st} value is 0.35, which is surprisingly high for a species with such a

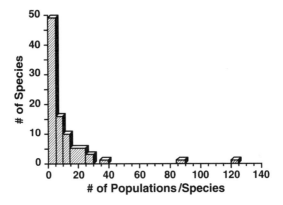

Figure 7.5. Frequency distribution of the number of extant populations for 91 plant species listed as threatened or endangered by the U.S. Fish and Wildlife Service (from Schemske et al., 1994. Redrawn with the permission of the Ecological Society of America).

narrow distribution. Gene flow between populations must be highly restricted. Most seedlings are established well within 0.5 meters of the parent plant.

Seeds were introduced into experimental sites 0.5 to 2.5 km away from extant populations. These sites are similar to those occupied by the species. Germination is as good as in natural sites; and seedling survival rates there are similar to those of seedlings that occurred in natural sites. Thus if seeds had been able to reach these sites, colonization may have occurred.

Colas et al. conclude that because populations are small and ostensibly undergo little gene exchange, they are subject to genetic drift, inbreeding depression, and mutational meltdown. Populations appear to lack colonizing ability so the species is dependent on the persistence of existing populations for its survival. These populations are at the mercy of stochastic environmental fluctuations, which can be very high. Population size appears to be declining, which places the species at the brink of extinction.

Chihuahua spruce (*Picea chihuahuana*) is another species whose long-term future is not favorable. The species was once widely distributed in central Mexico, but retreated northward during the Holocene warming. Ledig et al. (1997) reported that the species is known from only 35 stands in two northern states of Mexico. Most stands are small in size. A survey of 24 allozyme loci revealed that the species has a low level of genetic diversity (0.09), but that its populations are well differentiated ($F_{st} = 0.25$). Gene flow between populations is estimated at less than one individual per generation. Genetic drift apparently is involved, because there is no relationship between genetic distance and geographical distance and because the species occupies similar habitats in diverse parts of its range. On average 45% of the seeds are empty and seed germination is low, features which probably reflect inbreeding depression. The species is threatened by low seed production and inbreeding. Grazing and fire are likely to put additional pressure on population growth and restrict colonization of new sites. Some populations may be in an extinction vortex now.

Another species with a precarious future is *Geum radiatum* (Rosaceae), a perennial herb endemic to a few mountaintops in eastern Tennessee and western North Carolina. As reported by Hamrick and Godt (1996), 16 populations have been identified, but 5 have gone extinct. A sixth population may be in danger of extirpation from heavy foot traffic at a scenic overlook. Observations of four populations that currently are declining suggest that heavy recreational use may have played a prime role in recent extinctions. Five populations were analyzed for 25 allozyme loci. The overall genetic diversity is 0.10; 16% of its genetic diversity resides between populations. There is a close positive relationship between genetic diversity within populations and population size.

While the species mentioned above ostensibly represent examples of long-lived species threatened with extinction, let us turn our attention to younger species that are threatened. A prime example is Furbish's lousewort (*Pedicularis furbishiae*, Scrophulariaceae), a perennial herb. The species occurs in several localized populations in the St. John River Valley of Maine and adjacent New Brunswick, Canada. It is restricted to a limited, unstable, and ephemeral riverine habitat that is less than 12,000 years old (Menges, 1990). Catastrophic mortality of whole populations by riverine disturbance is common (Menges and Gawler, 1986). Local population extinction rates as high as 12% have been documented for some years. The annual founding rate of new populations has been 3% of viable populations. Even small annual probabilities of extinction result in almost certain extinction over moderate time scales. With 2% annual extinction, the 100-year probability of the survival of local lousewort populations is only 13% (Menges, 1990). The species depends on disturbance for the vitality of its populations. Too little or too frequent disturbance may cause the extirpation of local populations and the species as a whole. The species may lack the genetic variability to adapt to changes in its environment. A survey of 22 allozyme loci uncovered no variation (Waller, O'Malley, and Gawler, 1987).

A well-documented case of a threatened genus involves *Brighamia* (Campanulaceae), which is a halophytic, cliff-dweller endemic to Hawaii (Gemmill et al., 1998). The genus comprises two species, *B. insignis* from Kauai and Niihau and *B. rockii* from Molokai. The former in known from 29 plants on Kauai. All three of its populations have been declining rapidly, perhaps due to a virus and to herbivory by feral goats. One population declined from 12 to 4 plants between 1987 and 1990, then to 3 plants by 1992, and to 1 plant in 1996. A fourth population recently went extinct. The status of the species on the private island of Niihau is unknown. *Brighamia rockii* is estimated to contain only 170 individuals across its six known populations. Two populations have been declining rapidly. Seed-set from open pollination is poor in both species.

Allozyme diversity was surveyed in both species; and in both species gene diversity is a very low 0.06. Twenty-seven percent of the 11 loci scored are polymorphic in *B. insignis*, and 18% of the loci are polymorphic in *B. rockii*. Populations of *B. insignis* are weakly differentiated, $I = 0.95$. There is no variation among *B. rockii* populations, $I = 1.00$.

Extinction via Hybridization

The possibility that hybridization may lead to species extinction was first recognized by Harper et al. (1961), but only recently has its importance been appreciated (Rieseberg, 1991b; Ellstrand, 1992a; Levin, Francisco-Ortega, and Jansen, 1996). Rare insular species are particularly vulnerable to hybridization because of their frequent proximity to numerically larger congeners with which they may interbreed, and because of their small numbers. This threat is increasing because of the breakdown of ecological barriers, which affords species contact that otherwise may not have occurred. Insular congeners tend to have poorly differentiated floral architecture and unspecialized pollinators, both of which enhance the potential for hybridization.

The discussion of this topic will be divided into two sections. The first will deal with the processes involved in species extinction. The second will deal with examples.

The Process of Extinction

Interspecific hybridization fosters the extinction of populations by reducing the potential for plants to replace themselves, thereby inhibiting the growth of their populations. Population growth or contraction is determined by birth rate and death rate. Hybridization has a negative impact on the former.

The growth rate of a population may be retarded by the production of hybrid seed, which is made at the expense of conspecific seed. The consequence is the same whether hybrid seeds are viable or abort. When two species are intermixed and cross with equal facility as egg or pollen parents, a numerically small population will produce a greater proportion of hybrid seed than an abundant congener, as shown empirically in *Clarkia* (Lewis, 1961) and in analytical models (Levin, 1975; Fowler and Levin, 1984; Felber, 1991). The minority disadvantage increases as the crossing barriers between two species decline. If species have weak to moderate barriers, the minority species may fail to produce a sufficient quantity of seed to maintain its population size, and the population will shrink. Minority decline has been documented in hybrid swarms in *Senecio* (Murphy, 1981), *Helianthus* (Heiser, 1979), and *Carduus* (Warwick et al., 1989).

The numerical disadvantage of a minority species is compounded by the proliferation of fertile hybrids that are intercrossable with the rare species. The addition of these hybrids decreases the proportional representation of the minority species, which in turn causes a reduction in the proportion of their progeny that are conspecific.

Hybridization also reduces the growth rate of populations if hybrids successfully compete with parental types for establishment microsites and resources. The performance of "pure" plants may be suppressed more by F_1 hybrids than by conspecifics, as shown in *Anigozanthos* (Hopper, 1978), *Festuca* (O'Brien, Whittington, and Slack, 1967), and *Hordeum* (Norrington-Davies, 1972). The competitive status of advanced generation and backcross hybrids has yet to be investigated. Since most hybrids in a given population are not likely to be F_1s, the competitive status of other types of hybrids may be an important factor affecting the growth of the parental populations.

Hybridization also may reduce the growth rate of populations by increasing pathogen and herbivore pressure on them. This will occur when hybrids serve as a staging area for the colonization of the parental species. Hybrids often are more susceptible to pest exploitation than the parental species, and thus may support diverse and large pest arrays in close proximity to these species (Ericson, Burdon, and Wennstrom, 1993; Fritz, 1999; Whitham et al., 1999).

The genus *Spartina* provides an interesting example of how hybrids have a negative impact on one of the parental species. *Spartina alternifolia* was introduced into the San Francisco Bay, California, in the mid-1970s, where it coexists with the native *S. foliosa*. The species are interfertile, and hybrids are morphologically intermediate and vigorous. Hybrids are recruiting more rapidly than *S. foliosa* and threaten to replace this parent in San Francisco Bay (Daehler and Strong, 1997).

Examples of Hybridization and Extinction

The process of extinction via hybridization is well under way in *Cerocarpus traskiae* (Rosaceae), which occurs on the Santa Catalina Island off of southern California. The species is known from a single population. Morphological, allozyme, and RAPD analyses of seedlings and adult plants indicate that the species is hybridizing extensively with *C. betuloides*, which may absorb it (Rieseberg et al., 1989; Rieseberg and Gerber, 1995). Few pure *C. traskiae* plants remain. On San Clemente Island, also off the California coast, the endangered *Lotus scoparius* ssp. *traskiae* is threatened by hybridization with the abundant *L. argophyllus*. The process of assimilation is not as far along as in *Cerocarpus*.

Hybridization may threaten a species at different sites simultaneously, as in the composite *Argyranthemum coronopifolium* of Tenerife (Canary Islands). This rare species contains four "pure" populations and three populations in various stages of hybridization with the weedy and abundant *A. frutescens* (Brockmann, 1984; Bramwell, 1990). The latter has intruded into habitats of *A. coronopifolium* by migrating along corridors established by roadbuilding. Stages in the assimilation of *A. coronopifolium* by *A. frutescens* at "Risco del Fraile" have been documented over time. Species contact occurred in 1965 (Humphries, 1976b). In 1981 only a few morphologically "pure" *A. coronopifolium* occurred within a large hybrid swarm skewed toward *A. frutescens* (Brochmann, 1984). By 1996, the former was gone; there were only hybrids and *A. frutescens* (Levin et al., 1996). The assimilation of *A. coronopifolium* also is in progress at "Barranco del Rio," where only a few members of the species exist among many hybrids and numerous *A. frutescens*. The locations of hybridizing and pure populations of *A. coronopifolium* in relation to *A. frutescens* are depicted in figure 7.6.

The four pure populations of *A. coronopifolium* are not immediately threatened by their congener, but the small size of the species makes it very prone to extinction by other biotic or abiotic forces. The chance of species extinction is an inverse function of population number given that the extinction probabilities of separate populations are independent (Burgman, Ferson, and Akçakaya, 1993).

Other rare *Argyranthemum* species also are endangered by hybridization with *A. frutescens*. *Argyranthemum vincentii* is found in Tenerife in the northern pine forest as scattered plants, and in two populations several km away. One

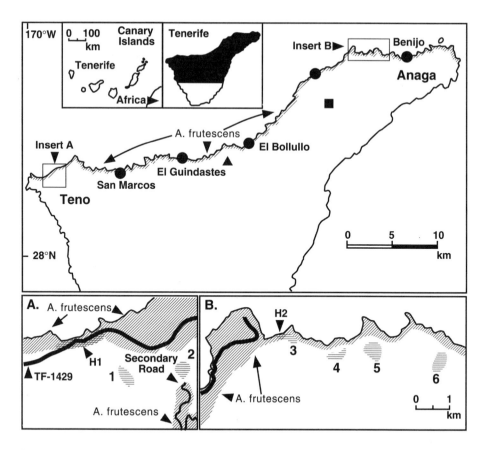

Figure 7.6. The locations of hybrid swarms (H1 and H2) and populations (1–6) of *Argyranthemum coronopifolium* in Tenerife. Population 2, although not overlapping *A. frutescens*, has hybrids in it.

population is hybridizing with *A. frutescens*, the other is not in contact with this species (Santos-Guerra, Francisco-Ortega, and Feria, 1993). *Argyranthemum sundingii* is known only from one population (12 flowering plants) embedded in lowland scrub on Tenerife (J. Francisco-Ortega, personal communication). The population is close to a road that extends to the coast, and along which *A. frutescens* has been migrating. The species recently was 2 km from *A. sundingii*, with which it readily hybridizes in the greenhouse.

Introduced weeds on islands may ultimately engulf congeneric natives. For example, in the Canary Islands, the introduced *Senecio vulgaris* hybridizes with the native *S. teneriffae* (Gilmer and Kadereit, 1989), and the introduced *Arbutus unedo* hybridizes with the rare *A. canariensis* (Salas-Pascual, Acebes-Gińoves, and Acro-Aguilar, 1993). The introduced *Gossypium barbadense* hybridizes with the endemic *G. tomentosum* in Hawaii (DeJoode and Wendel, 1992). In the British Isles, the introduced *Linaria repens* hybridizes with the rare *L. vulgaris*, and the introduced *Pinguicula grandiflora* hybridizes with the rare *P. vulgaris* (Stace, 1975).

The extinction of rare species also is progressing in continental floras. The rare Texas endemic *Hibiscus dasycalyx* is threatened by gene flow from *H. laevis* (Klips, 1995); and *Stellaria arenicola*, an endemic of the Athabasca sand dunes in Saskatchewan, Canada, is subject to gene flow from *S. longipes* (Purdy, Bayer, and Macdonald, 1994). Rhymer and Simberloff (1996) obtained a list of endangered species from the Nature Conservancy that are suspected of being threatened by hybridization and introgression based on morphological criteria. The list includes the last population of the Bakersfield saltbush (*Atriplex tularensis*), which is being assimilated by the widespread *A. serenana* in the Kern Lake Preserve in California. The California sycamore (*Platanus racemosa*), which occurs along the Sacramento River and its tributaries, is in the process of being amalgamated by the London plane (*P. acerifolia*).

In the previous examples and others often cited, a rare and numerically small species is threatened by hybridization with an abundant congener. The direction of gene flow is set by the numerical disparity between species. It is important to realize, however, that the challenger need not be numerically superior to the rare species to do considerable harm. This is demonstrated in the cordgrass *Spartina*, where a common native (*S. foliosa*) is locally threatened by a rare invader (*S. alterniflora*; Anttila et al., 1998). *Spartina alterniflora* was introduced into the San Francisco Bay, California, in the mid-1970s, where it coexists with the native *S. foliosa*. The invader produces five times as much pollen per anther as the native, and pollen of the invader germinates at approximately 2.5 times the rate of native pollen on native stigmas. Conspecific pollinations on the native yield 3 to 5% seed-set, whereas invader pollen on native stigmas yields about 23% seed-set. Thus the invader will sire the lion's share of seeds on plants of the native. Hybrids are vigorous and recruiting more rapidly than *S. foliosa* and threaten to replace this parent in San Francisco Bay (Daehler and Strong, 1997).

The extinction of a population or even a species by hybridization denotes the loss of a unique entity. It does not necessarily denote extinction of a local population or assemblage of populations. An introgressed entity may have greater vegetative and reproductive success than the original entity, and thus persist longer than its antecedent. Rattenbury (1962) made this point nearly four decades ago in his explanation of persistence and high levels of phenotypic variation in many tropical elements in the New Zealand flora.

Levels of Species Extinction

Periodically some species that are threatened or on the brink of extinction do go extinct. According to the World Conservation Monitoring Centre (1992), about 0.2% of angiosperm species and 0.3% of gymnosperm species have gone extinct over the past 400 years. Half of these extinctions occurred during this century (Smith et al. 1993). Roughly 27% of plant extinctions come from island endemics. This percentage far exceeds the percentage of species that are endemic to oceanic islands (17%; Heywood, 1979). The level of extinction varies among regions even though they may share similar climatic regimes.

Grueter (1995) observed that extinction levels in the floras with a Mediterranean climate in the Mediterranean region per se, Cape Province (South Africa), California, and Western Australia were 1.3, 3.0, 4.0, and 6.6%, respectively. The differences between regions correlate with the age and duration of human colonization. The lowest rate (in the Mediterranean proper) occurs where agriculture and grazing have been ongoing for 6000 to 8000 years, and the rate rises as the onset of European colonization is delayed.

There is considerable interest in plant extinction in the increasing number of tropical forest fragments. One notable study was on a 4-hectare isolated patch of lowland tropical rain forest preserved in the Singapore Botanical Gardens since 1859. Turner et al. (1996) compared a recent inventory with an extensive collection of herbarium specimens dating back to the 1890s. Of the 448 historically recorded species, 220 species remain. This amounts to a 51% extinction level. The degree of extinction differs among plant life forms. Herbs have lost 42% of their species, epiphytes, 67%, shrubs 74%, climbers 61%, and trees 42%. The failure of many species to recruit new individuals remains a matter of conjecture.

What can be expected of plant extinction within the next few decades? Species extinction rates may be estimated from species-area curves, which represent the relationship between the number of species in a region and its area. Using these curves, one may predict the proportion of species that will become extinct in a region based on the level of habitat loss. Raven (1987, 1988) and Myers (1988) have applied this technique to reach a global loss estimate of 7 to 9% of plant species per decade. Extinction rates are likely to be the highest in the tropics and subtropics because of deforestation there.

A second method of estimating future extinction rates is based on the rates at which species are "climbing the ladder" of the International Union for the Conservation of Nature and Natural Resources (IUCN) categories of threat, from vulnerable to endangered to probably extinct to certified extinction (Smith et al., 1993). Based on the differences in the number of species in the aforementioned categories between 1990 and 1992 and the amount given species have progressed, it is estimated that half of all higher plants would be extinct in 3,000 years, if the implied rate was accurate and sustained. Palms would be extinct in 50 to 100 years.

Both approaches to estimating future extinctions are based on very incomplete data bases (especially for most plant groups), the vagaries of sampling, misclassification of species in threat categories, and unknown spatial requirements for species persistence (May, Lawton, and Stork, 1995). Thus approximations from them are very crude.

Can we predict which species will go extinct? The species with smallest numbers of populations and whose populations are declining at the greatest rate are the prime candidates for extinction. Whereas there are estimates of population numbers and sizes for many rare species, we know almost nothing about their population size dynamics. The degree of rarity per se may not be a good predictor of the likelihood of extinction because some species may have traits that allowed them to become rare and persist in a rare state (Gaston and Kunin, 1997; Orians, 1997).

Traits that increase the proportion of potential pollen donors and recipients should be at a premium in rare species. In a genus where heterostyly is the ancestral condition, a shift to homostyly would facilitate mating in rare species and thus enhance their prospects of survival. The prediction that deviations from heterostyly are more likely in rare species than in common ones can be tested in the genus *Psychotria* (Rubiaceae), which contains more heterostylous species than any other angiosperm genus. Hamilton (1990) determined the ranges of 66 Mesoamerican species in this genus and found that homostyly was positively associated with rarity (table 7.4). Some rare species also were self-fertile.

Traits that reduce dependence on pollinators also should be at a premium in rare plants. Kunin and Schmida (1997) tested this expectation by relating the breeding systems of 52 Israeli crucifers to their abundance. Species with sparse populations are self-fertile more often than would be expected by chance. Moreover, rare self-incompatible species often have unusually large flowers, ostensibly to better attract pollinators. The association between rarity and self-fertility is by no means universal (Karron, 1987).

The Hawaiian endemic *Cyanea* (Campanulaceae) and its close ally *Rollandia* contain over 70 species that have very narrow geographical and elevational ranges, limited dispersal ability, and are dependent on a narrow range of bird pollinators (Givnish et al., 1995). Twenty-two percent of them are now extinct, and 29% are endangered. In *Cyanea* extinction has occurred exclusively among species with longer corolla tubes—that is, greater than 45 mm (Lammers, 1990). None of the more abundant species with tubes from 15 to 44 mm have suffered this fate (table 7.5). This suggests that the extinctions may be associated with the decimation of more specialized pollinators. The fate of the remaining long-tubed species may be in jeopardy.

Care must be taken in interpreting the cause of differences between rare and common species. Even if there were consistent differences between them, natural selection need not be the driving force (Orians, 1997). For example, the overrepresentation of self-fertility in rare species could result from the selective extinction of rare self-sterile plants, rather than selection for self-fertility per se.

Table 7.4. Floral morphology in relation to range size in *Psychotria* (from Hamilton, 1990).

Flower type	Range size[a]			
	Large	Medium	Small	Very Small
Homostylous	0	2	1	6
Heterostylous	14	8	15	4
Unknown	0	0	2	14

[a] based on number of 1° × 1° map squares

Table 7.5. Corolla tube lengths of extinct and
extant *Cyanea* species (based on Lammers, 1990).

Corolla tube length (mm.)	Number of Species	
	Extinct	Extant
15–29	0	10
30–44	0	13
45–59	7	4
60–74	5	7
75–89	1	2

May the spatial character of species rarity affect the threat of extinction? Is a species composed of a few large panmictic populations more vulnerable to the proximal causes of extinction than a series of small populations that collectively have the same number of individuals as few large populations? Although there is no simple answer, the risk of extinction probably is lower in the species with a few large populations (Lawton, 1995).

The Fate of Rare Species

There are five avenues that rare species may follow in their evolutionary histories. The one most written about and perhaps most traveled is extinction. The species is gone; the lineage terminates; there is nothing left of it. A second avenue that a species may travel is sustained existence. The species remains rare, perhaps being adapted for low densities. A third avenue is expansion. If the environment changes and creates suitable habitats that are within dispersal distance, a species may extend its range and eventually emerge from the ranks of the rare. It is likely that some species that are common today were rare at the height of the Wisconsin glaciation. A fourth avenue that a species may take is assimilation by another species through hybridization. Although the species disappears, part of its gene pool remains. All of these avenues have been discussed earlier in one form or another.

The fifth avenue open to rare species has not been discussed and rarely has been recognized by others (but see Stanley, 1979). In essence it is death and transfiguration. The species goes extinct, but the lineage does not. Some local population or group thereof respond to what Harlan Lewis (1962) calls "catastrophic selection" in ways that afford adaptations to an otherwise stressful environment. The surviving populations also may have incorporated novel genetic or chromosomal attributes via stochastic processes, in addition to those changes brought about by selection. The surviving populations, being ecologically distinct and perhaps partially reproductively isolated from the declining progenitor, may be considered a new species. Thus, while the vast majority of a terminal species' populations go extinct, some may persist in a new form. The demise of one species then would coincide with the origin of a second species. The replacement of one species by another in time is referred to as "anagenetic speciation." The replacement may occur relatively rapidly; it may appear to be punctuated.

A species with a few populations or many declining populations faces the same future as a series of isolated marginal populations of a well-distributed species. Both groups of populations inhabit sites to which they are ill-adapted or are deteriorating in quality. If the environment does not change, both groups are apt to go extinct. However, some population(s) within both groups may undergo a profound evolutionary change promoted by selection and stochastic processes that propel the novel populations to species status. There is one major difference between peripheral isolates of a viable species and a species near the brink of extinction. Speciation in peripheral isolates occurs without the demise of the progenitor, whereas anagenetic speciation accompanies the demise of the progenitor.

Even though a species has gone extinct, it may have allopolyploids derivatives. If these derivatives are very recent, the extinct genome may be more or less intact. There are many allopolyploids in which one parental species is extant and the other is unknown and apparently is extinct. For example, Roelfs et al. (1997) provide molecular evidence for an extinct common maternal parent of the tetraploids *Microseris acuminata* and *Microseris campestris*, but different "contemporary" paternal parents. Similarly, the maternal parent of the allopolyploid *M. scapigera* is extinct, whereas the paternal parent is a contemporary taxon or closely related to it (Van Houten, Scarlett, and Bachmann, 1993).

Overview

A multitude of plant species has narrow and/or fragmented distributions; they are classified as rare. Increasing awareness of the vulnerability of these species to extinction has prompted exploration of their genetics, demographics, and dispersal and population dynamics. Many studies have attempted to link rarity with species attributes. The connections often are rather tenuous.

Whereas some rare species are threatened with extinction, a significant proportion of rare species may not be facing this fate. We cannot infer the trajectory of a species (i.e., whether it is contracting, expanding or stationary) from the organization of its numbers in space at one point or a few points in time. Some rare species are neospecies in the process of expansion. Others are species expanding after a period of contraction.

Exclusive of human activities, the bases for species decline are poorly understood. Species ranges and population sizes certainly contract when there is climatic change. In most cases we do not know how changes in temperature and precipitation directly affect reproduction and survival, or indirectly affect these parameters through altered relationships with competitors, pests, and mutualists.

The fragmentation of large, continuous populations into many small discontinuous ones increases the ratio of perimeter to interior for populations. The consequences of such an increase have hardly been studied. Reproductive success in perimeter plants may differ from that in interior plants, as may the incidence of hybridization with neighboring populations.

The genetic compositions of small populations are affected more by a given number of immigrants than large populations. The impact of immigrants depends on their fitness in the recipient population. Although much is known about the performance of immigrants when introduced as seeds or transplants, little is known about their performance when introduced as pollen—that is, when coupled with another genome in hybrids. This is an important consideration, because in most species pollen typically is dispersed much farther than seeds.

On the one hand, gene flow may enrich local gene pools and thereby enhance the survivability of populations and ultimately species. On the other hand, gene flow may erode the genetic identity of populations and species if the source of genes is another species. Species generated during recent radiations on oceanic islands are particularly vulnerable to this fate.

8

The Persistence and Sorting of Incipient Species

The emergence of full-fledged neospecies is preceded by the evolution of incipient neospecies (species in *statu nascendi*). They have distinctive ecological profiles and/or genetic systems, but are not sufficiently divergent to warrant recognizing them as neospecies. Incipient neospecies will have short evolutionary futures if they cannot move beyond their locus of origin. In this chapter, I will discuss the fate of incipient neospecies, especially in terms of their survival and the impact that differential survival has on the character of full-fledged neospecies.

The Fate of Incipient Neospecies

Incipient species may evolve on the periphery of species' ranges or at distant (disjunct) sites. Mayr (1963) contends that peripheral isolates are formed on a regular basis in many species. Conversely, isolates located far from the periphery—that is, disjunct isolates—are likely to be formed much less frequently.

Peripatric and disjunct isolates have three possible fates. First, they can evolve into new species. The longer an isolate lives the greater its chance of becoming a full-fledged species. Second, they can persist without speciating. Third, they can become extinct. The latter is the most likely.

Mayr (1963) and Stanley (1979) contend that peripheral incipient neospecies face a very high rate of extinction during an extremely short period of geological time. Lewis (1962) observes that "most deviant marginal [*Clarkia*] populations are ephemeral and destined to become extinct *in situ*." Distant (disjunct) neospecies also are very vulnerable to extinction, as they are subjected to nearly the same pressures as peripheral neospecies, as discussed below. Thus most incipient neospecies are likely to be transient idiosyncrasies. Stanley (1978) refers to them as "aborted species."

The idea that derivatives are prone to extinction is supported chiefly by circumstantial evidence, because it is difficult or impossible to track derivatives over long time spans (Allmon, 1992). The best example of the extinction of one such entity is *Stephanomeria malheurensis*, which is derived from *S. exigua* ssp. *coronaria* (Gottlieb, 1973). Judging from the allozymic similarity of the taxa, *S. malheurensis* is of recent origin. The derivative was partially self-fertilizing, whereas its progenitor is an obligate outbreeder. The derivative was known only from a single hilltop south of Burns, Oregon, where it grew with its putative progenitor. It had an estimated population size of about 700 in the early to mid-1970s. It went extinct less than 20 years after it was first discovered, in part as the result of habitat changes and the invasion of *Bromus tectorum* (Guerrant, 1992).

Factors Promoting Incipient Species Extinction

The extinction of an incipient plant neospecies may be due to one or a combination of the following factors: (1) interspecific competition, (2) herbivory, (3) pathogens, (4) alteration of habitat, and (5) pollination limitation (MacArthur, 1972; Schemske et al., 1994). Genetic factors such as depauperate gene pools and inbreeding depression also might contribute to the demise of incipient neospecies (Barrett and Kohn, 1991; Ellstrand and Elam, 1993; Loeschcke, Tomiuk, and Jain, 1994). Both peripatric and disjunct incipient neospecies share the aforementioned environmental and genetical challenges. These challenges influence the vital rates of populations and ultimately species (i.e., birth, growth, and death) and have a negative impact on population size and successful expansion of the neospecies.

Ecological risk analyses of populations point to the vulnerability of incipient neospecies to extinction. The extinction risk is a function of the number of individuals it contains and the number of populations into which they are aggregated (Harrison and Quinn, 1989; Gilpin, 1990; Hanski, 1991; Gaston, 1990). Small local abundances and small geographical ranges make species especially vulnerable to environmental and demographic perturbations (Shaffer, 1987; Lande, 1988) and to systematic environmental change (Harrison, 1991; Thomas, 1994). The sooner a neospecies expands the numbers and sizes of its populations, the lower the probability of its extinction. The larger the range achieved, assuming a corresponding increase in population number and total number of individuals, the longer a new species is likely to persist (Fowler and MacMahon, 1982; Jablonski, 1987).

Local abundance is a prerequisite for the persistence of an outcrossing, incipient neospecies. If neospecies are wind-pollinated, they must occur in aggregations for normal seed-set because most pollen is deposited relatively close to the pollen source (Levin and Kerster, 1974). If plants are animal-pollinated, they also must occur in rather dense aggregates to reproduce normally. Otherwise, pollinator visitation rates will be low and pollinators will not be loyal to the species.

Self-incompatible specics that have been rare for long periods may evolve adaptations to overcome their numerical disadvantage. Relatively large and long-lived flowers would be examples of two such adaptations (Kunin and Shmida, 1997). However, we would not expect incipient neospecies to evolve adaptations for rarity. The time may be too short and genetic resources too sparse. If they were inadequately pollinated, they probably would go extinct.

Whereas many factors may cause the demise of incipient species, we know very little about the specifics of extinction in given situations. We need to identify those stages of the life cycle that are demographically most sensitive and the particular factors causing this sensitivity.

The Challenge of the Progenitor

All of the aforementioned factors being equal, peripatric and disjunct incipient species would have the same probabilities of survival and becoming full-fledged species. However, all else is not equal. The ebb and flow of species' borders may bring the incipient peripatric species into contact with its progenitor, which may be an agent of ecological and reproductive interference and of hybridization. In contrast, disjunct incipient neospecies are not subject to these challenges by their progenitors. As a result, incipient peripatric neospecies are likely to go extinct (for the reasons discussed below) at a higher rate than disjunct incipient species.

Reproductive Interference

The challenge of the progenitor has been considered almost exclusively in terms of competitive interactions (Mayr, 1963, 1982b; Stanley, 1979; Eldredge, 1989). However, progenitors also may impact the reproductive performance of their incipient neospecies through pollen exchange. Heterospecific pollen may interfere with conspecific pollen germination (Galen and Gregory, 1989), heterospecific microgametophytes may inhibit conspecific pollen tube growth (Gilissen and Linskens, 1975), and heterospecific microgametophytes may prevent the formation of conspecific seeds or cause their abortion (Williams et al., 1982; Harder, Cruzan, and Thomson, 1993). Thus the ability of incipient species to replace themselves may be hampered. The numerically superior progenitor can better tolerate the reduced seed production from these interactions.

The progenitor also may interfere with the reproductive process in its derivative through hybridization. Hybrid seeds are produced at the expense of conspecific seeds. The effect of reproductive interference on the survivorship of a derivative is well documented in *Clarkia* (Lewis, 1961). In mixed natural and artificial populations, the numerically superior progenitor *C. biloba* reproductively eliminates its derivative, *C. lingulata*. The latter produces an increasing proportion of hybrid seed as it becomes less numerous, and declines in population size because of its inability to replace itself. Hybrid sterility prevents the formation of hybrid swarms. *Clarkia lingulata* occurs alone in only two localities not far from its progenitor. Given that it has ecological requirements similar to its progenitor, the long-term persistence of the neospecies is unlikely.

The progenitor also may lead to the demise of the incipient neospecies by genetic swamping, if the latter is rare (Rieseberg, 1991b; Ellstrand and Elam, 1993; Levin et al., 1996). This process involves the formation of partially viable and partially fertile first-generation and backcross hybrids, and a growing number of hybrids. As hybrids increase in proportional representation, an increasing proportion of the seeds produced by a minority neospecies are hybrid. Eventually the neospecies is amalgamated by the more abundant progenitor. This process may be in progress in *Hibiscus*, where the peripheral endemic *H. dasycalyx* is threatened by gene flow from its progenitor *H. laevis* (Klips, 1995). Similarly, the narrow endemic *Stellaria arenicola* is subject to gene flow from its progenitor *S. longipes* (Purdy, Bayer, and Macdonald, 1994).

The inferences drawn from natural populations are supported by computer and analytical models. When hybrids are sterile or semisterile, crossing between an abundant species and a rare one often causes the latter's extinction (Crosby, 1970; Paterson, 1978; Lambert, Centner, and Paterson, 1984; Spencer, McArdle, and Lambert, 1986; Liou and Price, 1994). Even low levels of immigration into hybridizing populations from pure populations may fail to mitigate the demise of the rare species.

The level of reproductive interference experienced by the derivative depends in part on its breeding system. The most interference would accrue if it were an obligate outbreeder. The least would accrue if it were an obligate selfer, because the production of conspecific seed would not be affected by the identity of nearby pollen sources. Self-fertility is favored in peripheral neospecies derived from outcrossers because self-fertility is an isolating mechanism. Here are many examples of selfing taxa at the geographical margins of their outcrossing progenitors (Jain, 1976; Barrett and Kohn, 1991). The classic studies of peripatric speciation in *Clarkia* are a case in point (Lewis, 1962, 1973; Raven, 1964; Vasek, 1968; Bartholomew et al., 1973).

The aforementioned geographical relationship is not *prima facie* evidence that self-fertility arose to reduce reproductive interference. Self-fertility may have been favored for other reasons, such as the maintenance of adaptive gene complexes or the assurance of reproduction (Jain, 1976). However, selection for self-fertility in the face of hybridization has occurred in local populations of *Agrostis tenuis* and *Anthoxanthum odoratum* occupying mine sites with heavy metals in their soils (Antonovics, 1968). These populations are adjacent to conspecific populations that grow on normal pasture soils and that export pollen into the mine population. Hybrids are less tolerant of heavy metals than plants occupying the mine sites.

Ecological Interference

An incipient species may face considerable ecological interference as a result of contact with a numerically superior progenitor. Lewis (1966), Mayr (1963, 1982a & b), Stanley (1979), Eldredge (1989) and others contend that if peripheral derivatives have not acquired divergent habitat tolerances, they are apt to be overwhelmed by competition from an abundant progenitor.

There is almost no information on the competitive relationships between progenitors and their derivatives. One study of competition involves *Oenothera grandis* and its derivative, *Oe. laciniata* (Russell and Levin, 1988). The former is a bivalent former, whereas the latter is an obligate chromosomal heterozygote.

These species occasionally grow together in natural populations. Pairwise mixtures of the two species were grown in the field and in the greenhouse, and compared to monocultures grown in the same environments. There are no significant differences between the species in monocultures. However, the progenitor is the superior competitor in terms of relative yield and survivorship in mixtures.

Another study of competition involves the outcrossing *Stephanomeria exigua* and its derivative *S. malheurensis* (Gottlieb and Bennett, 1983). These species may have come into contact in nature. When grown in mixtures, the shoot biomass of *S. exigua* well exceeded that of its derivative even though the latter had greater biomass in pure stands. Thus the progenitor probably would eliminate its derivative should they intermix in nature.

Interference from Pests

Given that incipient species evolve in isolation from their progenitors, they may have escaped the pathogens, parasites, and herbivores of their progenitors. Should peripheral neospecies and progenitors make contact, neospecies populations would be subject to new biotic pressures because related host species often are susceptible to infection or attack by the same pathogens, parasites, and predators (Harlan, 1976; Fritz et al., 1994; Whitham et al., 1994). Pest pressure has been proposed to explain the maintenance of a parapatric distribution among related species that had the potential to invade each other's territory (Bull, 1991; Burger, 1992).

The exclusion of one species from the range of the other also may occur if both species share pests. The combined host density allows the mean level of infection to become higher than on a single host species, which may be more than one of the species can tolerate, or which may differentially affect the competitive abilities of the two species. Under this scenario, Price, Westoby, and Rice (1988) expect species with larger geographical ranges to displace those with smaller ranges. Incipient neospecies thus would be displaced by their progenitors.

Hybridization may facilitate the transfer of pests from one species to another. Floate and Whitham's (1993) "hybrid bridge hypothesis" predicts that intermediate hybrids allow pests restricted to one species gradually to adapt to another species. This hypothesis was suggested by the situation in *Populus*. Studies of eight species of leaf-galling species (one mite and seven aphid) show that in the absence of hybrids between *Populus fremontii* and *P. angustifolia* host shifting does not occur. However, in the presence of hybrids host shifting occurs. The presence of a spectrum of hybrids from advanced-generation backcrosses to first and advanced-generation hybrids may increase the potential for host shifts.

The transfer of fungal pathogens also may be dependent on the presence of hybrids, as seen in the floral smut fungus, *Anthracoidea fischeri*, on *Carex* (Ericson, Burdon, and Wennstrom, 1993). Where *C. canescens* and *C. mackenziei* occur in mixed populations but do not hybridize, both species are disease-free. However, where hybrids are present, both species and the hybrids are infected.

In summary, I propose that the challenge of the progenitor may threaten the survival of incipient neospecies. It does so through its effect on existing populations and by thwarting the expansion of the neospecies. The progenitor is not a problem faced by disjunct incipient species. These incipient species may establish large ranges and large populations before eventually making contact with the progenitor, if it indeed should transpire. Competition or hybridization at that point would not threaten its existence.

Extinction selectivity, or the relative vulnerability of species, is a topic of growing interest in ecological and paleontological circles, as recently reviewed by McKinney (1997). Properties often associated with increased extinction risk include specialization, low fecundity, limited mobility, and small geographical range. Discussions of extinction selectivity rarely deal with incipient species and do not include the challenge of the progenitor.

The Sorting of Neospecies

The survival or extinction of incipient neospecies depends on their genetic and phenotypic attributes. The differential survival of incipient neospecies may be referred to as "isolate selection" (Stanley, 1979). Isolate selection is among related isolates. Isolate selection differs from "catastrophic selection" (Lewis, 1962) in that the latter involves contractions in single populations that are accompanied by shifts in adaptive mode. Catastrophic selection is among individuals, whereas isolate selection is among incipient neospecies.

Selection among Isolates

Eldredge (1989) proposes that isolate selection favors entities whose ecological tolerances are divergent from those of their progenitors because this reduces the likelihood for competitive interactions with them. This proposal can be extended to the reproductive realm. I suggest that the differential survival of neospecies in the presence of the progenitor favors the accumulation of neospecies with relatively strong ecological and genomic barriers to hybridization with progenitors.

We can gain some insight into the possible if not actual operation of isolate selection by contrasting the genetic identities of peripheral neospecies and their putative progenitors with the strength of reproductive and or ecological isolation between them. Typically the genetic identity of neospecies and progenitors is high, in some cases approaching that of conspecific populations (Crawford, 1990). However, the level of ecological and genomic isolation between peripheral neospecies and progenitors is substantial, and may be stronger than the isolation between more distantly related congeners (e.g., *Clarkia*, Vasek, 1968; Lewis, 1973; *Coreopsis*, Crawford and Smith, 1982; *Crepis*, Togby, 1943; Sherman, 1946; *Gaura*, Raven and Gregory, 1972; Gottlieb and Pilz, 1976; *Gilia*, Grant, 1964; *Holocarpha*, Clausen, 1951; *Layia*, Gottlieb et al., 1985; *Lasthenia*, Crawford et al., 1985; *Mimulus*, Macnair, Macnair, and Martin, 1989). If the strength of ecological isolating barriers were a correlate of genomic divergence, we would expect neospecies to be weakly divergent from their progenitors.

The barriers between disjunct neospecies and progenitors tend to be moderate, even though the former may be the product of strong selection and genetic drift. As best we can tell, it is the species that evolved in distant localities that tend to hybridize most often when they make contact (Anderson, 1949; Remington, 1968; Rieseberg and Wendel, 1993). I contend that isolate selection, in addition to and perhaps rather than rapid speciation, has helped shape the barriers between peripheral neospecies and their progenitors.

Isolate selection would not strengthen existing ecological barriers between a neospecies and its progenitor because isolate selection does not act within populations. This process contrasts with the selective reinforcement of previously existing prezygotic barriers within populations, which is seen by some animal evolutionists as the final stage of speciation (e.g., Dobzhansky, 1940; Bock, 1972; Howard, 1993).

Whereas isolate selection favors neospecies that are ecologically and genomically divergent from their progenitors, it does not impose a directional bias on character expression, because the premium is on being different rather than on a specific character state. There may be several phenotypic, ecological, and genetical solutions to the challenge of the progenitor. Neospecies survival is not random; rather it is opportunistic. As a result, isolate selection is likely to produce highly autoapomorphic surviving neospecies, a state expected by Eldredge and Cracraft (1980), Wiley (1981), and Mayr (1982b) for peripheral neospecies.

The idea that differential survival of peripheral neospecies promotes divergence is consistent with the model of punctutated equilibrium, which proposes that most morphological change arises coincident with speciation or shortly thereafter (Eldredge and Gould, 1972). Proponents of this model argue that morphological changes result from stochastic processes (Gould, 1980). I have shown how morphological changes could result from selective processes as well.

Disjunct versus Peripatric Neospecies

Neospecies may evolve far from their ancestors as well as near them. These disjunct neospecies diverge from their ancestors as a result of local selective pressure and stochastic processes. Their long-term survival is not dependent on divergence from an ancestor. Therefore, disjunct neospecies are apt to be less divergent from an ancestor than a sister peripatric neospecies that began evolving at the same time.

To demonstrate whether the disjunct neospecies are less divergent than peripatric neospecies, one ideally compares the levels of divergence in many pairs (peripatric-disjunct) of neospecies derived from the same ancestor at the same time. Without knowledge of divergence time, one might compare sister neospecies that had similar genetic distances, but different spatial relationships with their common ancestor.

If disjunct neospecies are less divergent than peripatric neospecies, we might expect a correlation between genomic barriers and the geographical relationships of species. One example of this correlation is in the genus *Gilia* (Grant, 1966a). In the leafy-stemmed gilias, sympatric species have much stronger crossing barriers *inter se* than these species have with allopatric species or than allopatric species have *inter se*. In the cobwebby gilias, species with many sympatric contacts have stronger crossing barriers than species with few contacts. This pattern, although consistent with isolate selection, may have arisen in other ways, and thus does not constitute proof of its efficacy.

The premium on being different is greatest when incipient neospecies arise in sympatry with their progenitors. Populations composed of two species and hybrids occasionally spawn diploid hybrid species. The hybrid derivatives most divergent in reproductive traits and ecological tolerances have the best chance of expanding their ranges and becoming long-lived species. The least divergent ones are destined to become extinct through competition or hybridization with one or both species.

The best examples of stabilized diploid hybrid derivatives with divergent ecological requirements are in *Helianthus*. The parental species, *H. annuus* and *H. petiolaris*, are restricted to heavy clay soils, and to dry sandy soils, respectively (Rieseberg et al., 1995). Two hybrid derivatives, *H. anomalus* and *H. deserticola*, are endemic to habitats more xeric than tolerated by *H. petiolaris*. Another stabilized derivative of the same parentage, *H. paradoxus*, is narrowly distributed in brackish or saline marshes (Rieseberg et al., 1990). These recombinant types appear to be safe from amalgamation by their parents. Other recombinant types may have arisen only to be amalgamated by one of the parents because they had overlapping habitat requirements.

Polyploid species also arise in sympatry with their progenitor(s). Populations of autopolyploids occur sporadically throughout the range of many species (Lewis, 1980; Levin, 1983; Soltis and Soltis, 1993). When autopolyploidy is involved in speciation, divergence from progenitors through isolate selection is expected because polyploid neospecies will have been initiated in sympatry with their progenitor. Analytical models show that an emerging polyploid entity can be overwhelmed by a diploid by hybridization unless there are strong barriers to gene exchange (Fowler and Levin, 1984; Felber, 1991; Rodriguez, 1996). Evidence for the differential survivorship among autopolyploid populations comes from *Galax urceolata* in which diploids and autotetraploids hybridize readily (Nesom, 1983). Where the two cytotypes are sympatric, the autotetraploid occupies more mesic habitats; where they are allopatric this cytotype has very similar habitat preferences to its diploid relative.

I have argued that traits that reduce interspecific competitive and reproductive interference should be preferentially represented in successful neospecies. Such traits would allow neospecies' populations to grow and multiply. Glazier (1987) takes a broader view of isolate selection, and contends that any traits that prevent populations from declining or fluctuating widely in size would be preferentially represented in successful derivatives. Thus both peripatric and disjunct incipient species may be selected for self-fertility, long-lived seed banks, enhanced developmental plasticity, or some specific set of pheno-typic expressions which render a high degree of fitness.

The Phylogenetic Implications of Isolate Selection

If evolutionary differentiation occurs gradually and continuously over time with little or no contribution from speciation events, then the level of discrete character change will be similar in any pair of species relative to an outgroup, regardless of differences in the number of intervening speciation events (Avise and Ayala, 1975; Eldredge, 1989). This assumes that the pair of species are of equal age relative to the outgroup. If speciation contributes significantly to divergence, then species in lineages with many events will show more character change relative to an outgroup than species of equal age in a lineage with few events.

Given that more divergent peripatric neospecies have higher rates of success (expansion and longevity) than less divergent ones, the magnitude of phenotypic and ecological differentiation between species would be proportional to the number of successful speciation events since they last shared a common ancestor rather than to time elapsed. Correlatively, clades with predominantly peripatric speciation would be biased in favor of the accumulation of well-differentiated species (Lewis, 1966; Eldredge, 1989). Clades composed of species that evolved in distant allopatry would not be so biased because differential species survival would not be based on divergence from an ancestor.

There are no data on plants that demonstrate directly whether or not speciation promotes divergence. In a study on sunfish and minnows, Douglas and Avise (1982) showed that the magnitude of phenotypic divergence was independent of speciation rate. This is what would be expected of allopatric speciation. Mindell, Sites, and Grauer (1990) found a positive correlation between allozyme genetic distances and species richness in vertebrate genera that they attributed to the accelerating influence of speciation. However, the correlation may be spurious because of the numerous uncontrolled variables (Avise, 1994).

The accumulation of well-differentiated species following isolate selection would not necessarily be accompanied by the accumulation of species with large genetic distances because genetic distance between species primarily is a function of time (Nei, 1987). Accordingly, for a given genetic distance or divergence time, species of peripatric origin are expected to be more divergent (on average) from their progenitors than species of disjunct origin.

Overview

The origin of species is preceded by the origin of incipient species, species in *statu nascendi*. The proportion of incipient species that evolve into full-fledged, long-lived species ostensibly is very small. They almost invariably go extinct. Many reasons have been proposed for this failure, but little attention has been given to the challenge of the progenitor. The progenitor can lead to the demise of the incipient species through competition and hybridization, if the latter is on the periphery of the progenitors range.

The probability of a peripheral incipient species surviving is a positive function of the ecological and genomic disparities between it and its progenitor. In essence the progenitor "selects" divergent derivatives and "rejects" the others when the two are in contact. This process is not operative in the case of disjunct incipient species. Thus we might expect successful disjunct neospecies to be less divergent in their ecological and genomic relationships with their progenitors than peripatric neospecies.

Studies of the differential survival of lineages largely has been in the domain of macroevolution. We do ourselves a disservice if we ignore the same process at the microevolutionary level because it must be pervasive. There is a gap in our thinking about metapopulation dynamics (where species or groups thereof persist for different times), on the one hand, and the survival of lineages (where species or groups thereof persist for different times), on the other. The differential survival of lineages within species must have a profound effect on the long-term nature and fate of species.

9

Species Duration and the Tempo of Diversification

All species, regardless of their attributes or the character of the environment, will follow the passage from inception through expansion, differentiation, and demise. A multitude of plant species has completed this journey; and all extant species are now en route. Most contemporary species are either in the expansion or the differentiation phase. Before the massive impact of humans, the number of recent species probably exceeded the number in decline. Now the opposite may be the case.

Passage Time

The progression from birth through death is a property of species complexes as well as single species. Complexes begin with a single species, diversify through speciation, and ultimately go extinct. The passage time of single species will be considered first.

Single Species

The passage of a species may not simply involve expansion, differentiation, and demise. In response to major climatic change, species may undergo more than one cycle of expansion and contraction before they go extinct. This probably was the case during episodes of glacial expansion and retreat in North America and Europe. The genomic composition of species and patterns of geographical differentiation may vary among cycles, because the populations that survive one period of contraction may be rather different from those that survive another period (Hewitt, 1993). The distribution and abundance of species also may vary among cycles. Thus the history of a species may be far more complex than most of our studies suggest, especially if they are very long-lived.

The time to complete the passage will vary among species, because species differ in their intrinsic properties and in the temporal and spatial pattern of environmental change that they encounter. Longer species duration will be associated with high dispersal ability and broad habitat tolerance by and among individuals. The latter may be achieved by an open mating system that allows the accumulation of genetic variation within and among populations. Longer duration also will be associated with preference for wide-ranging resources. Therefore, I expect species with narrow soil preferences or those that depend on specialized pollinators to have relatively short lives.

The geographical location of species also may affect the passage time from inception to extinction. Species endemic to islands, especially oceanic islands, probably have relatively short lives. When faced with competition or environmental change, they may be unable to escape. As the level of competition is likely to decrease with the distance of islands from a major source of immigrants (MacArthur and Wilson, 1967), the mean longevity of island species should increase as the distance from a major source increases.

The longevity of species also may be negatively affected by their derivatives, if they are sympatric. New and superior forms may marginalize their predecessors while themselves becoming the focal point of future innovation. This pattern has been described in disparate groups of fossil animals (Pearson, 1998), but as yet not in plants.

The time to complete the passage may be approximated from the fossil record. Stanley (1975) argued that looking backward in geological time we will reach a point at which extant species comprise 50% of the flora. The average species duration will be twice this value. Levin and Wilson (1976) applied this estimator to plants. From paleobotanical data they reasoned that 50% of the late Pliocene herbs are extant, as are 50% of the late Miocene shrubs, 50% of the late Miocene hardwoods, and 50% of early Miocene conifers and cycads. Given these values, mean species duration times are 10 million years for herbs, 27 Myr for shrubs, 38 Myr for conifers, and 54 Myr for conifers and cycads. It is not surprising that herbs have much lower duration times than woody plants, given their penchant for greater variation in population size and shorter life spans.

The aforementioned estimates of duration times are considerably higher than those by Niklas, Tiffney, and Knoll (1983), which are based on the appearance and disappearance of species from the fossil record. They surmise that mean species durations within monocots and within dicots are about 3.5 million years and within gymnosperms are about 5.5 million years. They also surmise that within major plant groups species durations are short early in their histories and lengthen as the groups age.

The question yet to be addressed in plants is whether generalists have longer mean longevities than specialists as we might expect. This issue has been considered in animals. For example, in Osagean-Meramecian crinoids generalist clades have mean species longevities that are at least 45% greater than longevities in specialist clades (Kammer, Baumiller, and Ausich, 1998). Similar patterns of increased longevity in generalists have also been reported in marine gastropods (Gili and Martinell, 1994), marine bivalves (Stanley, 1986), shrews (Novacek, 1984), and antelopes (Vrba, 1987).

Estimates of mean species durations provide insights about mean extinction rates. As suggested by Stanley (1975), the mean rate of extinction is the reciprocal of the mean duration time. Using the estimates of Levin and Wilson (1976) above, herbs (with a duration of 10 Myr) have a mean extinction rate of $1/10$ or 0.10 species per million years, which signifies that 1 in 10 species will become extinct every million years. Shrubs have a mean extinction rate of 0.04 species per Myr, hardwoods 0.03 species per Myr, and conifers and cycads 0.02 species per Myr. Using the durations of Niklas et al. (1983), angiosperms have a mean extinction rate of 0.29 species per Myr and gymnosperms a mean of 0.18 per Myr.

The progression from neospecies through decline has elements in common with the taxon cycle described in several oceanic archipelagos (Ricklefs, 1989; Chown, 1997). In Stage I, a species inhabits an archipelago and expands its range, eventually colonizing all or most of the islands. In Stage II, populations on different islands differentiate to the level of subspecies. In Stage III, some of the differentiated taxa have gone extinct. In Stage IV, only one taxon remains on a single island. Eventually this taxon goes extinct. This cycle is thought to be driven by selection to exploit more stable and specialized habitats and by interspecific competition.

This cycle also may be thought of in terms of species divergence following colonization by a single progenitor. In stage III species would be formed on different islands, and in stage IV only a single species would remain. The cycle begins with species in disturbed, less-predictable environments and ends with species in more predictable environments.

The taxon cycle rarely has been considered of plants. One striking commentary is on the Hawaiian genus *Cyanea*. As described by Givnish (1998), the cycle begins with the occupation of open sites and low rates of speciation. It progresses with the invasion of closed habitats and high rates of speciation. Now it is in the extinction phase. Twenty-two percent of all species are extinct, and 29% are considered threatened. Extinction is concentrated in species with long specialized flowers and very narrow ranges.

The efficacy of the taxon cycle in plants also is supported by the character of species on continental islands. Smith (1981) has shown that the specialized and restricted alpine flora of Tasmania has evolved from generalist colonists with traits facilitating long-range dispersal.

Species Complexes

Stebbins (1971) recognized five stages in the history of polyploid complexes: (1) initial, (2) young, (3) mature, (4) declining, (5) relictual. In their initial stages, polyploid complexes are composed of wide-ranging diploid species and one or more narrowly distributed polyploids. In young complexes, some polyploids have overtaken and or surpassed the ranges of some diploids and are spreading aggressively. In mature complexes, the range of the diploids usually is narrower than that of the polyploids; and the diploids are allopatric and reproductively isolated. In declining complexes, some of the diploid progenitors have gone extinct so that some polyploids cannot be traced to extant parents. Ploidal levels also may increase. In relictual complexes, there are no diploids, and polyploids may have high ploidal levels.

Given that polyploid complexes have finite life spans, we may ask of factors that affect their longevity. There are five such factors.

1. The number of diploid and polyploid species. The more taxa in the complex, the longer that complex is likely to survive, all else being equal.
2. Range size. Larger ranges (and more populations) increase the longevity of each species and thereby the complex as a whole. Larger ranges also enhance the potential for additional bouts of hybridization and chromosome doubling.
3. The strength of prezygotic barriers. The weaker the barriers, the greater the potential for species to interbreed and generate new taxa.
4. The degree of ecological differentiation among species. More divergent species are less likely to succumb to the same agents of decline than are species interacting with the environment in similar ways.

The number of species, range size, strength of prezygotic barriers among species, and level of ecological differentiation among species also affect the longevity of other types of species complexes, such as the homogamic, heterogamic, and agamic complexes (Grant, 1981). A homogamic complex includes species and their hybrid derivatives at the same ploidal level. *Helianthus annuus, H. petiolaris*, and their three hybrid derivatives constitute a homogamic complex. A heterogamic complex includes species with normal bivalent pairing and hybrid derivatives that are obligate chromosomal heterozygotes. The genus *Oenothera* is especially well known for such complexes (Cleland, 1972). An agamic complex includes sexually reproducing species and hybrid derivatives that are partially or completely asexually reproducing. Some species are likely to be polyploid. These complexes occur widely in such remote groups as the Asteraceae, Rosaceae, and Poaceae (Grant, 1981).

Having noted that species complexes rise and fall at a rate dependent on several factors, it is useful to compare relative life expectancies of different species complexes. A case may be made that life expectancies are likely to vary in consonance with the ability of their components to exchange genetic material and to generate new hybrid derivatives, all else being equal. Grant (1981) argues that the derived components of heterogamic and agamic complexes may be cut off from their progenitors and other components as well. Their recombination systems are closed or nearly so, thus retarding the genesis of new adaptations and new taxa. This is likely to be detrimental to the longevity of the derived components and thereby to the longevity of the complex as a whole.

The derived components within homogamic and polyploid complexes would have the ability to hybridize with their parental entities or other derived components. Hybridization between derived components could yield new derivatives or enrich the gene pools of the existing components. The recombination system of homogamic derivatives largely is open, while that of polyploids is only partially restricted. The ability to generate new taxa is especially important to the longevity of complexes when the diploid pillars go extinct.

How long a complex may persist is a matter of conjecture, as is the age of a given complex. According to Stebbins (1971), distributional evidence suggests that some agamic complexes are millions of years old. He cites related New World and Old World complexes in the blackberries (*Rubus* subg. *Eubatus*), that probably arose and spread along with the Arcto-Tertiary flora during the Tertiary period. He also suggests that agamic complexes in *Pennisetum, Panicum*, and *Setaria* are of Tertiary origin. Stebbins also proposes, from distributional patterns, that most mature polyploid complexes are between 500,000 and 10 million years old.

Rates of Speciation

We have seen how the fossil record can be used to gain insights into the longevities and extinction rates of groups of species. The same record in conjunction with knowledge on contemporary diversity within genera or broader taxa can be used to gain insight into rates of speciation. Levin and Wilson (1976) estimated the net number of speciation events (r_s) per Myr for groups of genera using the following equation:

$$r_s = \frac{3.3}{N} \sum_{i=1}^{N} \left[\frac{\log s}{t} \right)$$

where N is the number of genera, s is the number of living species in a genus and t is the age of the genus. This equation is similar to that used by Stanley (1975) to estimate speciation rates in single genera. This is the simple exponential equation for population growth: $N = N_o\, e^{Rt}$, where N is the number of species, N_o the initial number of species (one for strict monophyly), t is the time, and R is the intrinsic rate of increase in number of species. It assumes an exponential increase in the number of lineages per genus through time—that is, a dichotomously branching phylogeny. The net speciation rate is the true speciation rate minus the extinction rate.

The mean net speciation rates per Myr for different growth forms are presented in table 9.1. The speciation rate in herbs is the highest, two orders of magnitude above that for cycads. The mean speciation rate for the three groups of flowering plants is 0.46. This value is not far removed from the 0.37 estimated for angiosperms by Niklas et al. (1983). The estimates for conifers and cycads in table 9.1 are an order of magnitude higher than those of Niklas et al. (1983).

It is of interest to apply Stanley's approach to other groups of plants such as those on island archipelagos. If we know the number of single-island species per island, the total number of native species on each island, and the age of each island, we can estimate the mean speciation rate by solving for R. These estimates will be conservative because all progenitors did not arrive soon after

Table 9.1. Estimates of speciation rate parameters (per Myr) from Levin and Wilson, 1976).

Growth form	Mean species duration time in Myr(D)	Mean species extinction rate (1/D)	Net speciation rate	True speciation rate
Herbs	10	0.10	1.05	1.15
Shrubs	27	0.04	0.24	0.28
Hardwoods	38	0.03	0.09	0.12
Conifers	54	0.02	0.02	0.04
Cycads	54	0.02	0.01	0.03

the emergence of the island. The requisite information has been gathered by Sakai et al. (1995) for Hawaiian Island angiosperms. Species numbers, island ages, and the estimates of speciation rates per lineage per Myr on each island are presented in table 9.2. Speciation rates are highest on the youngest island, Hawaii (2.1 species per lineage per Myr). The mean rate is 0.80 species per lineage per Myr. Speciation rates are independent of island size.

Total and endemic species numbers for the genus *Clermontia* in the Hawaiian Islands are provided by Lammers (1995). The speciation rate on Hawaii (max. age 0.5 Myr) is estimated at 6.3 per lineage per Myr versus 0.6 on Maui (max. age 1.8 Myr) and 0.3 on Oahu (max. age 3.8 Myr).

The Juan Fernandez Islands comprise Mastierra (ca. 4 Myr) and Masafuera (1–2 Myr; Stuessy et al., 1984). The former has 55 endemic plus 26 native species, and the latter has 35 endemic plus 15 native species (Stuessy et al., 1998). The estimated rate of speciation on Mastierra is 0.33 species per lineage per Myr; the estimated rate for Masafuera is 0.96 species per lineage per Myr. As in the Hawaiian Islands, speciation rates are highest on the youngest island.

If speciation rates are a function of ecological opportunity or ecological persistence, then we would expect speciation rates to be higher on oceanic islands than on continents. This is what Schluter (1998) found when he compared Hawaiian clades of silverswords, *Tetramolopium* and *Bidens*, with their probable mainland sister clades. The rate differential between oceanic islands also is quite apparent in animals as well.

The idea that speciation rates depend on ecological opportunity was extended by Ehrlich and Raven (1969) to sets of interacting species. They pro-

Table 9.2. Net speciation rates in the Hawaiian islands.

	Kauai	Oahu	Maui Nui	Hawaii
Total endemic species	350	345	432	286
Single-island endemics	163	124	158	79
Island age (Myr)	5.7	3.8	1.8	0.5
Speciation rate per Myr	.20	.28	.60	2.1

Species numbers and island ages are from Sakai et al. (1995).

posed that plant lineages diversify at a greater rate when they are freed from herbivores through the development of a novel defense. This hypothesis was tested by Farrell, Dussourd, and Mitter (1991), who compared the diversities of lineages that have independently evolved resin and latex canal systems with sister groups in 16 plant lineages. They found that speciation rates in clades with secretory canals are significantly higher than rates in sister taxa that lack secretory canals.

Nectar spurs may have been another such key innovation. The evolution of nectar spurs would allow a shift to a more specialized pollinator fauna. Does this shift affect speciation rates? Studies on the spur-bearing *Aquilegia* and its close allies, the spurless *Isopyrum* and *Thalictrum*, suggest that this may be the case (Hodges and Arnold, 1994). Hodges (1997) compared species numbers in eight clades with floral spurs and their spurless sister groups, and found that seven had more species in the clades with spurs.

Stanley's estimator of speciation rates may be applied to archipelagos as a whole, as well as to single genera. This is desirable when the distances between islands is small, because species spread from their islands of origin. Consider the Canary Islands, a group of seven islands. The archipelago contains 1,262 angiosperm species of which 201 are endemic (Rodriguez and Perez de Paz, 1997). Approximately 65% of the species are native (J. Francisco-Ortega, personal communication). The mean age of the islands is ca. 12 million years (Francisco-Ortega et al., 1997). Using the estimated number of native species (820), the net rate of speciation per lineage is 0.09 per Myr.

From estimates of the net speciation rate and the mean extinction rate, we may easily obtain an estimate of the true speciation rate per Myr by adding the two values together. This is provided in table 9.1. The true speciation rate in herbs is about an order of magnitude greater than that of hardwoods, which in turn is almost an order of magnitude greater than that in cycads (Levin and Wilson, 1976). What is the relationship between the tempo of speciation and the life span of species? In herbs and shrubs, the average speciation event is roughly 10% of the average life span of a species. In hardwoods it is about 3%, and in gymnosperms it is less than 1%.

The true speciation rates and extinction rates among plant growth forms in table 9.1 are highly correlated, $r = 0.93$. A lower correlation between these rates ($r = 0.47$) has been described among major groups of land plants (Niklas et al., 1983). Stanley (1990) also reported a significant correlation for these rates among major animal phylads. He contends that such correlations reflect common causal factors that elevate or depress both rates simultaneously. It seems that speciation and extinction rates are both negatively correlated with niche width, and with population size and stability. Habitat fragmentation may promote speciation and extinction. Dispersal ability may be inversely correlated with speciation and extinction rates in species with relatively stable and continuous distributions, whereas dispersal ability may be positively correlated with these rates in species with unstable and patchy distributions.

Speciation in plants often is associated with chromosomal change. Whereas it is impossible to measure chromosomal change that has no bearing on chromosome number, it is possible to estimates rates of chromosome number evolution from the heterogeneity of these numbers within genera. Levin and Wilson (1976) obtained numbers for 8,378 angiosperm and 590 gymnosperm species from the literature. They first estimated the mean increase in chromosome number diversity along a typical lineage within a genus. This quantity, m, is given by $m = (k - 1)/nt$, where n is the number of species sampled per genus and k is the number of chromosome numbers per genus. By dividing m by the putative age of the genus (t), the mean rate of increase in chromosomal diversity per lineage is obtained. The following equation was used to estimate the average rate of increase in chromosomal diversity per lineage (r_c) for a group of N genera:

$$r_c = \frac{1}{N} \sum_{i=1}^{N} \frac{(k - 1)}{nt}$$

This equation provides a minimum estimate because all species with the same derived chromosome number are considered to have evolved from a single common ancestor with that number.

Herbs have the highest mean rate of chromosome evolution for both polyploidy and aneuploidy (table 9.3). Surprisingly, the rate was an order of magnitude higher in shrubs than in hardwoods. Gymnosperms display little chromosomal change.

The rate of chromosomal diversification is significantly correlated with the rate of species diversification among angiosperm genera, $r = 0.64$. When genera are grouped into growth forms, some striking correlations emerge between rates of chromosomal and species diversification: $r = 0.71$ for herbs, 0.89 for shrubs, and 0.15 for trees. The absence of a significant correlation in trees is not readily explained.

The general correlation among angiosperms ostensibly is due to the regulation of both processes by common factors including the size of the breeding unit and incidence of bottlenecks in the history of genera. Localized breeding and repeated colonization-extinction episodes promote both speciation and chromosomal evolution.

It is unlikely that the average rates of increase in chromosome number and species number diversity per lineage have been constant through the history of a genus. Phylads usually evolve much more rapidly early in their history than

Table 9.3. Net rates of chromosomal evolution versus growth form (from Levin and Wilson, 1976).

| Growth form | Mean increase per lineage per Myr | | |
	Total	Polyploid	Aneuploid
Herbs	0.073	0.050	0.020
Shrubs	0.010	0.010	0.0005
Hardwoods	0.0013	0.001	0.0003
Conifers	0.0001	0.00001	0.0001
Cycads	0.0000	0.0000	0.0000

in more mature stages (Stebbins, 1974; Stanley, 1979). For example, rates of increase in species and chromosomal diversity in woody genera of the Magnoliidae probably were much greater during the Upper Cretaceous and Paleocene than during subsequent periods (Ehrendorfer, 1968; Stebbins, 1971).

Finally, we may consider speciation rates in relation to community type. We must approach this issue indirectly because we don't have the information. As argued earlier, by virtue of their expansion, successful new species are invaders. They spread from their locales of origin. Therefore, the rate of successful speciation in a lineage depends on the ability of the evolutionary novelty to expand, to invade a community.

Are all communities equally invasible? Apparently not. Communities with depauperate species diversity and life-form diversity are more invasible than those rich in these respects (Elton, 1958; Rejmánek, 1989). Communities with disturbance or long-term change are more invasible than communities without it (Crawley, 1987; Rejmánek, 1989). Early successional communities are more invasible than late successional or climax communities (Johnstone, 1986; Rejmánek, 1989). Continental tropical communities, especially forests, are more resistant to invasion that temperate communities (Rejmánek, 1996).

If there is a relationship between invasibility and successful speciation, then speciation rates should be the highest in communities depauperate in species and life forms. This is indeed the case. As noted earlier in this chapter, younger oceanic islands have higher speciation rates than older islands, and the latter have higher speciation rates than continents as a whole. Whether speciation rates are relatively high in weedy species and in communities undergoing high rates of disturbance or change remains to be determined.

Having considered possible speciation rate differences among communities, it is interesting to speculate on speciation rate differences among angiosperm families as suggested by the taxonomic pattern of plant invasions. Pysek (1998) surveyed the alien floras of 26 regions distributed over the globe and covering a broad variety of habitats. For our purposes, we are interested in the families that are overrepresented relative to the species pool available as potential invaders. The families with unusual invasion ability are the Chenopodiaceae, Cruciferae, Papaveraceae, Poaceae, and Polygonaceae. Therefore, these families may have relatively high speciation rates. Whether they do remains to be determined. The Asteraceae and Leguminosae contribute large numbers of aliens to local floras, but they are not overrepresented.

The differences in speciation rates across communities and families would be most likely to occur if the ability to mobilize and fix evolutionary novelty were the same in all plant groups, ecological and taxonomic. We don't know if this is the case. Nevertheless, looking at relative speciation rates from the invasion perspective allows the formulation of hypotheses that may be testable some day.

Overview

The time between the birth and death of species is unknown, but no doubt varies widely as a result of differences in the properties of species, the spatial and temporal pattern of environmental exigencies, and serendipity. The mean species duration for herbs may be as much as 10 million years and 38 million years for hardwoods. The rate of extinction is negatively correlated with species duration as is the rate of speciation. Rates of chromosomal and species diversification are positively correlated. Speciation rates appear to the highest where ecological opportunity is high and competition is low, as is the case on young islands. Key innovations also are associated with high speciation rates.

Species complexes arise and die. They persist longer than individual species, how much so depending on the level of ecological and geographical diversification and on the number of species and their overall population sizes. The maturation of polyploid complexes often is accompanied by a decline in the range size and number of the diploid pillars. In some complexes no diploids are extant. Some complexes may be millions of years old.

The focal point of this book has been change—change in the genetic attributes and ranges of species through time. The temporal genetic variable is the least understood of the two because it can only be addressed indirectly when the time span is centuries or longer. Distributional change is tractable in wind-pollinated species (and to lesser extent others) through the use of fossil pollen.

The temporal genetic variable is tractable on a short time scale. However, studies of this variable have been indirect, using phenotypic variation as a surrogate for genetic variation. For example, Stebbins and Daly (1961) monitored changes in a hybrid swarm between *Helianthus annuus* and *H. bolanderi* over an eight-year period. Heiser (1979) described changes in three hybrid swarms involving *Helianthus divaricatus* and *H. microcephalus* over a 22-year period. Murphy (1981) returned to a hybrid swarm involving *Senecio bicolor* and *S. jacobaea* that had been described 78 years earlier. Warwick et al. (1989) reanalyzed a hybrid zone involving *Carduus nutans* and *C. acanthoides*, which had been described at various intervals over a 30-year period.

Some temporal studies are independent of hybridization, and they too demonstrate change. For example, New (1978) reanalyzed clines in presence or absence of seedcoat papillae and in hairiness of stems and leaves in *Spergula arvensis* after 20 years. Epling, Lewis, and Ball (1960) studied the proportion of blue- and white-flowered plants in several populations of *Linanthus parryae* over 13 years.

Genetical or ecological studies of short duration do not contribute to an understanding of long-term changes during the passage of species. However, they do highlight the need to bring a temporal perspective to bear whenever possible. Expanding young systems like the *Tragopogon* polyploids of the northwestern United States are not the only venue for studying change in time. For example, studies on the rare Chihuahua spruce, as discussed in chapter 8, combine contemporary understanding of genetic variation and spatial distribution with a fossil record of the Holocene history of the species. Contemporary data alone afford only one entry in a tabulation of rare species variation.

Literature Cited

Abbo, S., and G. Ladizinsky. 1994. Genetical aspects of hybrid embryo abortion in the genus *Lens* L. Heredity 72: 193–200.

Abbott, R. J., and J. A. Irwin. 1988. Pollinator movements and the polymorphism for outcrossing rate at the ray locus in Groundsel, *Senecio vulgaris* L. Heredity 60: 295–298.

Ågren, J. 1996. Population size, pollinator limitation, and seed set in the self-incompatible herb *Lythrum salicaria*. Ecology 77: 1779–1790.

Aitken, S. N., and W. J. Libby. 1994. Evolution of pygmy-forest edaphic subspecies of *Pinus contorta* across an ecological staircase. Evolution 48: 1009–1019.

Aizen, M. A., and P. Feinsinger. 1994a. Forest fragmentation, pollination, and plant reproduction in a chaco dry forest, Argentina. Ecology 75: 330–351.

Aizen, M. A., and P. Feinsinger. 1994b. Habitat fragmentation, native insect pollinators, and feral honey bees in Argentine "Chaco Serrano." Ecol. Applic. 4: 378–392.

Alexander, D. E. 1960. Performance of genetically induced corn tetraploids. American Seed Trade Association 15: 68–74.

Al-hiyaly, S. A., T. McNeilly, and A. D. Bradshaw. 1988. The effects of zinc contamination from electricity pylons—evolution in a replicate situation. New Phytol. 110: 571–580.

Allee, W. C. 1949. Group survival value for *Philodina roseola*, a rotifer. Ecology 30: 395–397.

Allen, G. A., J. A. Antos, A. C. Worley, T. A. Suttill, and R. J. Hebda. 1995. Morphological and genetic variation in disjunct populations of the avalanche lily *Erythronium montanum*. Canad. J. Bot. 74: 403–412.

Allendorf, F. W., K. L. Knudsen, and G. M. Blake. 1982. Frequencies of null alleles at enzyme loci in natural populations of ponderosa and red pine. Genetics 100: 497–504.

Allendorf, F. W., and R. F. Leary. 1986. Heterozygosity and fitness in natural populations of animals. Pages 57–76 in M. Soulé, ed., Conservation Biology: The Science of Scarcity and Diversity. Sinauer Associates, Sunderland, Mass.

Allmon, W. D. 1992. A causal analysis of stages in allopatric speciation. Oxford Surv. Evol. Biol. 9: 219–257.

Anderson, E. 1949. Introgressive Hybridization. John Wiley & Sons, New York.

Anderson, E., and G. L. Stebbins. 1954. Hybridization as an evolutionary stimulus. Evolution 8: 378–388.

Andersson, L. 1990. The driving force: species concepts and ecology. Taxon 39: 375–382.

Anttila, C. K., C. C. Dahler, N. E. Rank, and D. R. Strong. 1998. Greater male fitness of a rare invader (*Spartina alternifolia*, Poaceae) threatens a common native (*Spartina foliosa*) with hybridization. Amer. J. Bot. 85: 1597–1601.

Antrobus, S., and A. J. Lack. 1993. Genetics of colonizing and established populations of *Primula veris*. Heredity 71: 252–258.

Antonovics, J. 1968. Evolution in closely adjacent plant populations. V. Evolution of self-fertility. Heredity 23: 219–238.

Antonovics, J. 1976. The nature of limits to natural selection. Ann. Missouri Bot. Gard. 63: 224–247.

Armbruster, W. S. 1993. Evolution of plant-pollination systems: hypotheses and tests with the neotropical vine *Dalechampia*. Evolution 47: 1480–1505.

Armbruster, W. S., and K. E. Schwaegerle. 1996. Causes of covariation of phenotypic traits among populations. J. Evol. Biol. 9: 261–276.

Arnold, M. L. 1997. Natural Hybridization and Evolution. Oxford University Press, New York.

Ashton, P. A., and R. J. Abbott. 1992. Multiple origins and genetic diversity in the newly arisen allopolyploid species, *Senecio cambrensis* Rosser (Compositae). Heredity: 68: 25–32.

Auld, B. A., J. Hosking, and R. E. McFadyen. 1983. Analysis of the spread of tiger pear and parthenium weed in Australia. Australian Weeds 2: 56–60.

Avise, J. C. 1994. Molecular Markers, Natural History and Evolution. Chapman & Hall, New York.

Avise, J. C., and F. J. Ayala. 1975. Genetic change and rates of cladogenesis. Genetics 81: 757–773.

Avise, J. C., and R. M. Ball, Jr. 1990. Principles of genealogical concordance in species concepts and biological taxonomy. Oxford Surv. Evol. Biol. 7: 45–67.

Ayres, M. P. 1993. Plant defense, herbivory, and climate change. Pages 75–94 in P. M. Kareiva, J. G. Kingsolver, and R. B. Huey, eds., Biotic Interactions and Global Change. Sinauer, Sunderland, Mass.

Baker, A. M., and J. S. Shore. 1995. Pollen competition in *Turnera ulmifolia* (Turneraceae). Amer. J. Bot. 82: 717–725.

Baker, H. G. 1965. Characteristics and modes of origins of weeds. Pages 147–168 in H. G. Baker and G. L. Stebbins, eds., The Genetics of Colonizing Species. Academic Press, New York.

Baker, H. G. 1972. Migrations of weeds. Pages 327–347 in D. H. Valentine, ed., Taxonomy, Phytogeography, and Evolution. Academic Press, New York.

Baker, H. G. 1974. The evolution of weeds. Ann. Rev. Ecol. Syst. 5: 1–24.

Baker, H. G. 1986. Patterns of plant invasion in North America. Pages 44–57 in H. A. Mooney and J. A. Drake, eds., Ecology of Biological Invasions in North America and Hawaii. Springer-Verlag, New York.

Baker, H. G., and P. D. J. Hurd. 1968. Intrafloral ecology. Ann. Rev. Entomol. 13: 385–414.

Baldwin, B. G. 1997. Adaptive radiation of the Hawaiian silversword alliance: congruence and conflict of phylogenetic evidence from molecular and non-molecular investigations. Pages 103–128 in T. J. Givnish and K. J. Sytsma, eds., Molecular Evolution and Adaptive Radiation. Cambridge University Press, Cambridge.

Baldwin, B. G., and R. H. Robichaux. 1995. Historical biogeography and ecology of the Hawaiian silversword alliance (Asteraceae)–new molecular and phylogenetic perspectives. Pages 259–285 in W. L. Wagner and V. A. Funk, eds., Hawaiian Biogeography-Evolution on a Hot Spot Archipelago. Smithsonian Institution Press, Washington.

Bannister, M. H. 1965. Variation in the breeding system of *Pinus radiata*. Pages 353–372 in H. G. Baker and G. L. Stebbins, eds., The Genetics of Colonizing Species. Academic Press, New York.

Barigozzi, C. ed. 1982. Mechanisms of Speciation. Alan R. Liss, Inc. New York.

Barrett, S. C. H., and D. Charlesworth. 1991. Effects of a change in the level of inbreeding on the genetic load. Nature 352: 522–524.

Barrett, S. C. H., L. D. Harder, and A. C. Worley. 1996. The comparative biology of pollination and mating in flowering plants. Phil. Trans. Royal Soc. Lond. B. 351: 1271–1280.

Barrett, S. C. H., and B. C. Husband. 1990. The genetics of plant migration and colonization. Pages 254–277 in A. H. D. Brown, M. T. Clegg, A. L. Kahler, and B. S. Weir, eds., Plant Population Genetics, Breeding, and Genetic Resources. Sinauer Associates, Sunderland, Mass.

Barrett, S. C. H., and J. R. Kohn. 1991. Genetic and evolutionary consequences of small population size in plants: implications for conservation. Pages 3–30 in D. A. Falk and K. E. Holsinger, eds., Genetics and Conservation of Rare Plants. Oxford University Press, New York.

Barrett, S. C. H., M. T. Morgan, and B. C. Husband. 1989. The dissolution of a complex genetic polymorphism: the evolution of self-fertilization in tristylous *Eichhornia paniculata* (Pontederiaceae). Evolution 43: 1398–1416.

Barrett, S. C. H., and B. J. Richardson. 1986. Genetic attributes of invading species. Pages 21–33 in R. H. Groves and J. J. Burdon, eds., Ecology of Biological Invasions: An Australian Perspective. Australian Academy of Science, Canberra.

Bartholomew, M., L. C. Eaton, and P. H. Raven. 1973. *Clarkia rubicunda*: a model of plant evolution in semi-arid regions. Evolution 27: 505–517.

Barton, N. H. 1979. The dynamics of hybrid zones. Heredity 43: 341–359.

Barton, N. H. 1995. Linkage and the limits to natural selection. Genetics 140: 821–841.

Barton, N. H., and B. Charlesworth. 1984. Genetic revolutions, founder effects, and speciation. Ann. Rev. Ecol. Syst. 15: 133–164.

Barton, N. H., and G. M. Hewitt. 1985. Analysis of hybrid zones. Ann. Rev. Ecol. Syst. 16: 113–148.

Bateman, A. J. 1947. Contamination of seed crops. I. Insect pollination. J. Genet. 48: 257–275.

Bauert, M. R., M. Kalin, M. Baltisberger, and P. J. Edwards. 1998. No genetic variation detected within isolated relict populations of *Saxifraga cernua* in the Alps using RAPD markers. Molec. Ecol. 7: 1519–1527.

Baum, D. A., and M. J. Donoghue. 1995. Choosing among alternative "phylogenetic" species concepts. Syst. Bot. 20: 560–573.

Baum, D. A., and K. L. Shaw. 1995. Genealogical perspectives on the species problem. Pages 289–303 in P. C. Hoch and A. G. Stephenson, eds., Experimental and Molecular Approaches to the Plant Biosystematics. Missouri Botanical Garden, St. Louis.

Bazzaz, F. A., D. A. Levin, and M. R. Schmierbach. 1982. Differential survival of genetic variants in crowded populations of *Phlox*. J. Appl. Ecol. 19: 891–900.

Beltran, I. C., and S. H. James. 1974. Complex hybridity in *Isotoma petraea*. IV. Heterosis in interpopulation hybrids. Austral. J. Bot. 22: 251–264.

Bennington, C. C., and J. B. McGraw. 1996. Environment-dependence of quantitative genetic parameters in *Impatiens pallida*. Evolution 50: 1083–1097.

Berhan, A. M., S. H. Hulbert, L. G. Butler, and J. L. Bennetzen. 1993. Structure and evolution of the genomes of *Sorghum bicolor* and *Zea mays*. Theor. Appl. Genet. 86: 598–604.

Bernacchi, D., and S. D. Tanksley. 1997. An interspecific backcross of *Lycopersicon esculentum* × *L. hirsutum*: linkage analysis and a QTL study of sexual compatibility factors and floral traits. Genetics 147: 861–877.

Berry, P. E., and R. N. Calvo. 1994. An overview of the reproductive biology of *Espeletia* (Asteraceae) in the Venezuelan Andes. Pages 229–249 in R. W. Rundel, A. P. Smith, and F. C. Meinzer, eds., Tropical Alpine Environments. Cambridge University Press, Cambridge.

Bhiry, N., and N. Filion. 1996. Mid-Holocene hemlock decline in eastern North America linked with phytophagous insect activity. Quaternary Res. 45: 312–320.

Bijlsma, R., N. J. Ouborg, and R. van Treuren. 1994. On genetic erosion and population extinction in plants: a case study of *Scabiosa columbaria* and *Salvia pratensis*. Pages 255–271 in V. Loeschcke, J. Tomiuk, and S. K. Jain, eds., Conservation Genetics. Birkhäuser Verlag, Basel, Switzerland.

Billington, H. L., A. M. Mortimer, and T. McNeilly. 1988. Divergence and genetic structure in adjacent grass populations. I. Quantitative genetics. Evolution 42: 1267–1277.

Birks, H. J. B. 1989. Holocene isochrone maps and patterns of tree-spreading in the British Isles. J. Biogeogr. 16: 503–540.

Birky, C. W. 1988. Evolution and variation in plant chloroplast and mitochondrial genomes. Pages 23–53 in L. D. Gottlieb and S. K. Jain, eds., Plant Evolutionary Biology, Chapman & Hall, London.

Birky, C. W., T. Maruyama, and T. Fuerst. 1983. An approach to population and genetic theory for genes in mitochondria and chloroplasts, and some results. Genetics 103: 513–527.

Bixby, P. J., and D. A. Levin. 1996. Response to selection for autogamy in *Phlox*. Evolution 50: 892–899.

Bock, W. J. 1972. Species interactions and macroevolution. Evolutionary Biology 5: 1–24.

Bond, W. J. 1995. Assessing the risk of plant extinction due to pollinator and disperser failure. Pages 131–146 in J. H. Lawton and R. M. May, eds., Extinction Rates. Oxford University Press, Oxford.

Bonierdale, M. W., R. L. Plaisted, and S. D. Tanksley. 1988. RFLP maps based on a common set of clones reveal modes of chromosomal evolution in potato and tomato. Genetics 120: 1095–1103.

Bossart, J. L., and D. P. Prowell. 1998. Genetic estimates of population structure and gene flow: limitations, lessons and new directions. Trends Ecol. Evol. 13: 202–205.

Boudry, P., M. Morchen, P. Sanmitou-Laprade, P. Vernet, and H. Van Dijk. 1993. The origin and evolution of weed beets: consequences for the breeding and release of herbicide resistant transgenic sugar beets. Theor. Appl. Genet. 87: 471–478.

Bradshaw, A. D. 1984. The importance of evolutionary ideas in ecology—and visa versa. Pages 1–25 in B. Shorrocks, ed., Evolutionary Ecology, Blackwell Scientific Publ., London.

Bradshaw, A. D. 1991. Genostasis and the limits to evolution. Phil. Trans. Royal Soc. Lond. B. 333: 289–305.

Bradshaw, A. D., and T. McNeilly. 1981. Evolution and Pollution. Edward Arnold, London.

Bradshaw, H. D., S. M. Wilbert, K. G. Otto, and D. W. Schemske. 1995. Genetic mapping of flower traits associated with reproductive isolation in monkeyflowers (*Mimulus*). Nature 376: 762–765.

Bramwell, D. 1990. Conserving biodiversity in the Canary Islands. Ann. Missouri Bot. Gard. 77: 28–37.

Brochmann, C. 1984. Hybridization and distribution of *Argyranthemum coronopifolium* (Asteraceae-Anthemideae) in the Canary Islands. Nordic J. Bot. 4: 729–736.

Brooks, D. R., and D. A. McLennan. 1991. Phylogeny, Ecology, and Behavior. University of Chicago Press, Chicago.

Brown, A. H. D., and J. J. Burdon. 1987. Mating systems and the colonizing success of plants. Pages 115–131 in A. J. Gray, M. J. Crawley, and P. J. Edwards, eds., Colonization, Succession and Stability. Blackwell Scientific Publ., Oxford.

Brown, A. H. D., and D. R. Marshall. 1981. Evolutionary changes accompanying colonization in plants. Pages 351–364 in G. C. E. Scudder and J. L. Reveal, eds., Evolution Today, Proc. II Intl. Congress Syst. Evol. Biol.

Brown, J. H., G. C. Stevens, and D. M. Kaufman. 1996. The geographic range: size, shape, and internal structure. Ann. Rev. Ecol. Syst. 27: 597–623.

Brown, W. L., Jr., and E. O. Wilson. 1956. Character displacement. Syst. Zool. 5: 49–64.

Broyles, S. B. 1998. Postglacial migration and the loss of allozyme variation in northern populations of *Asclepias exaltata* (Asclepiadaceae). Amer. J. Bot. 85: 1091–1097.

Broyles, S. B., A. Schnabel, and R. Wyatt. 1994. Evidence for long-distance pollen dispersal in milkweed (*Asclepias exaltata*). Evolution 48: 1032–1040.

Bull, C. M. 1991. Ecology of parapatric distributions. Ann. Rev. Ecol. Syst. 22: 19–36.

Bruneau, A. 1997. Evolution and homology of bird pollination syndromes in *Erythrina* (Leguminoseae). Amer. J. Bot. 84: 54–71.

Brunken, J., J. M. J. de Wet, and J. R. Harlan. 1977. The morphology and domestication of pearl millet. Econ. Bot. 31: 163–174.

Burdon, J. J., D. R. Marshall., and A. H. D. Brown. 1983. Demographic and genetic changes in populations of *Echium plantagineum*. J. Ecol. 71: 667–679.

Burger, R., and M. Lynch. 1995. Evolution and extinction in a changing environment: a quantitative-genetic analysis. Evolution 49: 151–163.

Burger, W. 1992. Parapatric close-congeners in Costa Rica: hypotheses for pathogen-mediated plant distribution and speciation. Biotropica 24: 567–570.

Burgman, M. A., S. Ferson, and H. R. Akçakaya. 1993. Risk Assessment in Conservation Biology. Chapman & Hall, London.

Bush, E. J., and S. C. H. Barrett. 1993. Genetics of mine invasions by *Deschampsia caespitosa*. Canad. J. Bot. 71: 1336–1348.

Byers, D. L. 1998. Effect of cross proximity on progeny fitness in a rare and common species of *Eupatorium* (Asteraceae). Amer. J. Bot. 85: 644–653.

Byers, D. L., and T. R. Meagher. 1992. Mate availability in small populations of plant species with homomorphic sporophytic self-incompatibility. Heredity 68: 353–359.

Cain, M. L., H. Damman, and A. Muir. 1998. Seed dispersal and the Holocene migration of woodland herbs. Ecol. Monogr. 68: 325–347.

Campbell, D. R. 1996. Evolution of floral traits in a hermaphroditic plant: field measurements of heritabilities and genetic correlations. Evolution 50: 1442–1453.

Campbell, D. R., N. M. Waser, M. V. Price, E. A. Lynch, and R. J. Mitchell. 1991. Components of phenotypic selection: pollen export and flower corolla width in *Ipomopsis aggregata*. Evolution 45: 1458–1467.

Campbell, D. R., N. M. Waser, and E. J. Meléndez-Ackerman. 1997. Analyzing pollinator-mediated selection in a plant hybrid zone: hummingbird visitation patterns on three spatial scales. Amer. Natur. 149: 295–315.

Campbell, D. R., N. M. Waser, and M. V. Price. 1996. Mechanisms of bird-mediated selection for flower width in *Ipomopsis aggregata*. Ecology 77: 1463–1472.

Carey, K. 1983. Breeding system, genetic variability, and response to selection in *Plectritis* (Valerianaceae). Evolution 37: 947–956.

Carlquist, S. 1974. Island Biology. Columbia University Press, New York.

Carlquist, S. 1995. Introduction. Pages 1–13 in W. L. Wagner and V. A. Funk, eds., Hawaiian Biogeography—Evolution on a Hot Spot Archipelago. Smithsonian Institution Press, Washington.

Carlson, P. S., H. H. Smith, and R. O. Dearing. 1972. Parasexual interspecific plant hybridization. Proc. Natl. Acad. Sci. USA. 69: 2292–2294.

Carr, G. D. 1976. Chromosome evolution and aneuploid reduction in *Calycadenia pauciflora* (Asteraceae). Evolution 29: 681–699.

Carr, G. D. 1977. A cytological conspectus of the genus *Calycadenia* (Asteraceae), an example of contrasting modes of evolution. Amer. J. Bot. 64: 694–703.

Carr, G. D. 1980. Experimental evidence for saltational chromosome evolution in *Calycadenia pauciflora* Gray. Heredity 45: 109–115.

Carr, G. D. 1998. A fully fertile intergeneric hybrid derivative from *Argyroxiphium sandwicense* ssp. *macrocephalum* × *Dubautia menziesii* (Asteraceae) and its relevance to plant evolution in the Hawaiian Islands. Amer. J. Bot. 82: 1574–1581.

Carr, G. D., and D. W. Kyhos. 1981. Adaptive radiation in the Hawaiian silversword alliance (Compositae: Madiinae). Evolution 35: 543–556.

Carr, G. D., and D. W. Kyhos. 1986. Adaptive radiation in the Hawaiian silversword alliance (Compositae: Madiinae). II. Cytogenetics of artificial and natural hybrids. Evolution 40: 959–976.

Carr, R. L., and G. D. Carr. 1983. Chromosome races and structural heterozygosity in *Calycadenia ciliosa* (Asteraceae). Amer. J. Bot. 70: 744–755.

Carson, H. L. 1968. The population flush and its genetic consequences. Pages 123–137 in R. C. Lewontin, ed., Population Biology and Evolution. Syracuse University Press. Syracuse.

Carson, H. L., and A. R. Templeton. 1984. Genetic revolutions in relation to speciation phenomena: the founding of new populations. Ann. Rev. Ecol. Syst. 15: 97–131.

Carter, R. N., and S. D. Prince. 1981. Epidemic models to explain biogeographic distribution limits. Nature 293: 644–645.

Carter, R. N., and S. D. Prince. 1988. Distribution limits from a demographic viewpoint. Pages 165–184 in A. J. Davy, M. J. Hutchings, and A. R. Watkinson, eds., Plant Population Ecology. Blackwell Scientific Publ., London.

Carey, K. 1983. Breeding system, genetic variability, and response to selection in *Plectritis* (Valerianaceae). Evolution 37: 947–956.

Carey, P. D., A. R. Watkinson, and F. F. O. Gerard. 1995. The determinants of the distribution and abundance of the winter annual grass *Vulpia ciliata* ssp. *ambigua*. J. Ecol. 83: 177–187.

Case, M. A. 1993. High levels of variation within *Cypripedium calceolus* (Orchidaceae) and low levels of divergence among its varieties. Syst. Bot. 18: 663–677.

Chandler, J. M., C. Jan, and B. H. Beard. 1986. Chromosomal differentiation among the annual *Helianthus* species. Syst. Bot. 11: 353–371.

Charlesworth, B. 1992. Evolutionary rates in partially self-fertilizing species. Amer. Natur. 140: 126–148.

Charlesworth, B., R. Lande, and M. Slatkin. 1982. A neo-Darwinian commentary on macroevolution. Evolution 36: 474–498.

Charlesworth, B., M. T. Morgan, and D. Charlesworth. 1991. Multilocus models of inbreeding depression with synergistic selection and partial self-fertilization. Genet. Res. 57: 177–194.

Charlesworth, D., and B. Charlesworth. 1987. Inbreeding depression and its evolutionary consequences. Ann. Rev. Ecol. Syst. 18: 237–268.

Charlesworth, D., and B. Charlesworth. 1990. Inbreeding depression with heterozygote advantage and its effect on selection for modifiers changing the outcrossing rate. Evolution 44: 870–888.

Charlesworth, D., M. T. Morgan, and B. Charlesworth. 1990. Inbreeding depression, genetic load, and the evolution of outcrossing rates in a multilocus system with no linkage. Evolution 44: 1469–1489.

Chown, S. L. 1997. Speciation and rarity: separating cause from consequence. Pages 91–109 in W. E. Kunin and K. J. Gaston, eds., The Biology of Rarity. Chapman & Hall, London.

Chu, Y. N., and H. I. Oka. 1972. The distribution and effects of genes causing F_1 weakness on *Oryza breviligulata* and *O. glaberrima*. Genetics 70: 163–173.

Clark, J. S., C. Fastie, G. Hurtt, S. T. Jackson, C. Johnson, G. A. King, M. Lewis, J. Lynch, S. Pacala, C. Prentice, E. W. Schupp, T. Webb III, and P. Wyckoff. 1998. Reid's paradox of rapid plant migration. BioScience 48: 13–24.

Clausen, J. 1951. Stages in the Evolution of Plant Species. Cornell University Press, Ithaca.

Clausen, J., and W. M. Hiesey. 1958. Experimental studies on the nature of species. IV. Genetic structure of ecological races. Carnegie Institution of Washington. Publ. No. 615.

Clausen, J., D. D. Keck, and W. M. Hiesey. 1940. Experimental studies on the nature of species. I. Effects of varied environments on western American plants. Carnegie Institution of Washington Publ. No. 520.

Clausen, J., D. D. Keck, and W. M. Hiesey. 1947. Heredity of geographically and ecologically isolated races. Amer. Natur. 81: 114–133.

Cleland, R. E. 1972. *Oenothera*: Cytogenetics and Evolution. Academic Press, New York.

Coates, D. J. 1988. Genetic diversity and population genetic structure in the rare Chittering grass wattle *Acacia anomala* Court. Austral. J. Bot. 36: 273–286.

Coates, D. J. 1992. Genetic consequences of a bottleneck and spatial genetic structure in the triggerplant *Stylidium coroniforme* (Stylidiaceae). Heredity 69: 512–520.

Cocks, P. S., and J. R. Phillips. 1969. Evolution of subterranean clover in South Australia. I. The strains and their distribution. Austral. J. Agric. Res. 30: 1035–1052.

Cohan, F. M., and A. A. Hoffmann. 1989. Unifying selection as a diversifying force in evolution: evidence from *Drosophila*. Amer. Natur. 134: 613–637.

Colas, B., I. Olivieri, and M. Riba. 1997. *Centaurea corymbosa*, a cliff-dwelling species tottering on the brink of extinction: a demographic and genetic study. Proc. Natl. Acad. Sci. USA. 94: 3471–3476.

Comes, H. P. 1998. Major gene effects during weed evolution: phenotypic characters cosegregate with alleles at the ray floret locus in *Senecio vulgaris* L. (Asteraceae). J. Hered. 89: 54–61.

Comps, B., B. Thiébaut, L. Paule, D. Merzeau, and J. Letouzey. 1990. Allozyme variability in beechwoods (*Fagus sylvatica* L.) over central Europe: spatial differentiation among and within populations. Heredity 65: 407–417.

Copes, D. L. 1981. Isozyme uniformity in western red cedar seedlings from Oregon and Washington. Canad. J. For. Res. 11: 451–453.

Cousins, R., and M. Mortimer. 1995. Dynamics of Weed Populations. Cambridge University Press, New York.

Couvet, D., F. Bonnemaisson, and P-H. Gouyon. 1986. The maintenance of females among hermaphrodites: the importance of nucleocytoplasmic interactions. Heredity 57: 325–330.

Cowling, R. M., and P. M. Holmes. 1992. Endemism and speciation in a lowland flora from the Cape Floristic Region. Biol. J. Linn. Soc. 47: 367–383.

Coyne, J. A. 1992. Genetics and speciation. Nature 355: 511–515.

Coyne, J. A. 1995. Speciation in monkeyflowers. Nature 376: 726–727.

Coyne, J. A., and R. Lande. 1985. The genetic basis of species differences in plants. Amer. Natur. 126: 141–145.

Cracraft, J. 1983. Species concepts and speciation analysis. Curr. Ornith. 1: 159–187.

Crawford, D. J. 1979. Allozyme variation in *Chenopodium incanum*: Intraspecific variation and comparison with *Chenopodium fremontii*. Bull. Torrey Bot. Club 106: 257–261.

Crawford, D. J. 1990. Plant Molecular Systematics. John Wiley & Sons. New York.

Crawford, D. J., and R. J. Bayer. 1981. Allozyme divergence in *Coreopsis cyclocarpa* (Compositae). Syst. Bot. 6: 373–386.

Crawford, D. J., R. Ornduff, and M. C. Vasey. 1985. Allozyme variation within and between *Lasthenia minor* and its derivative species *L. maritima* (Asteraceae). Amer. J. Bot. 72: 1177–1184.

Crawford, D. J., and E. B. Smith. 1982. Allozyme variation in *Coreopsis nuecensoides* and *C. nuecensis*: a progenitor-derivative species pair. Evolution 36: 379–386.

Crawford, D. J., and E. B. Smith. 1984. Allozyme divergence and intraspecific variation in *Coreopsis grandiflora*. Syst. Bot. 9: 219–225.

Crawford, D. J., and T. F. Stuessy. 1998. Plant speciation on oceanic islands. Pages 249–267 in K. Iwatsuki and P. H. Raven, eds., Evolution and Diversification in Land Plants. Springer-Verlag, Tokyo.

Crawford, D. J., T. F. Stuessy, and M. Silva. 1987. Allozyme divergence and the evolution of *Dendroseris* (Compositae: Lactuceae) on the Juan Fernandez Islands. Syst. Bot. 12: 435–443.

Crawford, D. J., R. Witkus, and T. F. Stuessy. 1987. Plant evolution and speciation on oceanic islands. Pages 183–199 in K. M. Urbanska, ed., Differentiation Patterns in Higher Plants. Academic Press, London.

Crawford, R. M. M. 1989. Studies in Plant Survival. Blackwell, Oxford.

Crawley, M. J. 1987. What makes a community invasible? Pages 429–453 in A. J. Gray, M. J. Crawley, and P. J. Edwards, eds., Colonization, Succession and Stability. Blackwell, Oxford, England.

Critchfield, W. B. 1984. Impact of the Pleistocene on the genetic structure of North American conifers. Pages 70–118 in R. M. Lanner, ed., Proceedings of the Eighth North American Forest Biology Workshop. Utah State University, Logan.

Cronn, R., M. Brothers, K. Klier, P. K. Bretting, and J. F. Wendel. 1977. Allozyme variation in domesticated annual sunflower and its wild relatives. Theor. Appl. Genet. 95: 532–545.

Crosby, J. L. 1970. The evolution of genetic discontinuity: computer models of the selection of barriers to interbreeding between subspecies. Heredity 25: 253–297.

Croullebois, M. L., M. T. Barreneche, H. de Cherisey, and J. Pernes. 1989. Intraspecific differentiation of *Setaria italica* (L.) P. B.: study of abnormalities (weakness, segregation distortion, and partial sterility) observed in F_1 and F_2 generations. Genome 32: 203–207.

Crow, J. F., W. R. Engels, and C. Denniston. 1990. Phase three of Wright's shifting balance theory. Evolution 44: 233–247.

Cwynar, L. C., and G. M. MacDonald. 1987. Geographical variation of lodgepole pine in relation to population history. Amer. Natur. 129: 463–469.

Daehler, C. C., and D. R. Strong. 1993. Predictions and biological invasions. Trends Ecol. Evol. 8: 380.

Daehler, C. C., and D. R. Strong. 1997. Hybridization between introduced smooth cordgrass (*Spartina alternifolia*; Poaceae) and native California cordgrass (*S. foliosa*) in San Francisco Bay, California, USA. Amer. J. Bot. 84: 607–611.

Dana, M. N., and P. D. Ascher. 1985. Pseudo-self-compatibility (PSC) in *Petunia integrifolia*. J. Hered. 76: 468–470.

Dancik, B. P., and F.C Yeh. 1983. Allozyme variability and evolution of lodgepole pine (*Pinus contorta* var. *latifolia*) and jack pine (*Pinus banksiana*) in Alberta. Canad. J. Genet. Cytol. 25: 57–64.

Darley-Hill, S., and W. Johnson. 1981. Acorn dispersal by the Blue Jay (*Cyanocitta cristata*). Oecologia 50: 231–232.

Darlington, C. D., and K. Mather. 1949. Elements of Genetics. Allen and Unwin, London.

Davis, M. B. 1981. Outbreaks of pathogens in Quaternary history. IV International Palynology Conference, Lucknow (1976–1977) 3: 216–228.

Davis, M. B. 1983. Holocene vegetational history of the eastern United States. Pages 166–181 in H. E. Wright, ed., Late-Quaternary Environments of the United States. Vol. 2. The Holocene. University of Minnesota Press, Minneapolis.

Davis, M. B. 1987. Invasions of forest communities during the Holocene: beech and hemlock in the Great Lakes region. Pages 373–393 in A. J. Gray, M. J. Crawley, and P. J. Edwards, eds., Colonization, Succession, and Stability. Blackwell Scientific Publ., London.

Davis, M. B. 1989. Insights from paleoecology on global climatic change. Bull. Ecol. Soc. Amer. 70: 222–228.

DeJoode, D. R., and J. F. Wendel. 1992. Genetic diversity and origin of the Hawaiian Islands cotton, *Gossypium tomentosum*. Amer. J. Bot. 79: 1311–1319.

Delcourt, P. A., and H. R. Delcourt. 1987. Long-term Forest Dynamics of the Temperate Zone. Springer-Verlag, New York.

DeMauro, M. 1991. Relationship of breeding system to rarity in the lakeside daisy (*Hymenoxys acaulis* var. *glabra*). Conserv. Biol. 7: 542–550.

Demesure, B., B. Comps, and R. J. Petit. 1996. Chloroplast DNA phylogeography of the common beech (*Fagus sylvatica* L.) in Europe. Evolution 50: 2515–2520.

de Nettancourt, D. 1977. Incompatibility in Angiosperms. Springer-Verlag, New York.

dePamphlis, C. W., and R. Wyatt. 1990. Electrophoretic confirmation of interspecific hybridization in *Aesculus* (Hippocastanaceae) and the genetic structure of a broad hybrid zone. Evolution 44: 1295–1317.

Derks, F. H. M., J. C. Hakkert, W. H. J. Verbeek, and C. M. Colijn-Hooymans. 1992. Genome composition of asymmetric hybrids in relation to the phylogenetic distance between parents. Nucleus-chloroplast interaction. Theor. Appl. Genet. 84: 930–940.

deVicente, M. C., and S. D. Tanksley. 1993. QTL analysis of transgressive segregation in an interspecific tomato cross. Genetics 134: 585–596.

Devlin, B., and N. C. Ellstrand. 1990. The development and application of a refined method for estimating gene flow from angiosperm paternity analysis. Evolution 44: 248–259.

Dewey, D. R. 1966. Inbreeding depression in diploid, tetraploid, and hexaploid crested wheatgrass. Crop Sci. 6: 144–147.

Dewey, D. R. 1969. Inbreeding depression in diploid and induced autotetraploid crested wheatgrass. Crop Sci. 9: 592–595.

Dobzhansky, Th. 1940. Speciation as a stage in evolutionary divergence. Amer. Natur. 74: 312–321.

Dobzhansky, Th. 1951. Genetics and the Origin of Species, 3rd ed. Columbia University Press, New York.

Doebley, J. 1989. Molecular evidence for a missing wild relative of maize and the introgression of its chloroplast genome into *Zea perennis*. Evolution 43: 1555–1559.

Doebley, J., and R.-L. Wang. 1997. Genetics and the evolution of plant form: an example from maize. Cold Spring Harbor Symp. Quant. Biol. 62: 361–367.

Dolan, R. W. 1994. Patterns of isozyme variation in relation to population size, isolation, and phytogeographic history in royal catchfly (*Silene regia*; Caryophyllaceae). Amer. J. Bot. 81: 965–972.

Donoghue, M. J., and P. D. Cantino. 1988. Paraphyly, ancestors, and the goals of taxonomy: a botanical defense of cladism. Bot. Rev. 54: 107–128.

Douglas, M. E., and J. C. Avise. 1982. Speciation rates and morphological divergence in fish: tests of gradual versus rectangular modes of evolutionary change. Evolution 36: 224–232.

Dubcovsky, J., M-C Luo, G-Y Zhong, R. Bransteitter, A. Desai, A. Kilian, A. Kleinhofs, and J. Dvorak. 1996. Genetic map of diploid wheat, *Triticum monococcum* L., and its comparison with maps of *Hordeum vulgare* L. Genetics 143: 983–999.

Ducousso, A., H. Michaud, and R. Lumaret. 1993. Reproduction and gene flow in the genus *Quercus* L. Ann. Sci. For. Suppl. 1 (Paris) 50: 91s-106s.

Dudash, M. R. 1990. Relative fitness of selfed and outcrossed progeny in a self-compatible, protandrous species, *Sabatia angularis* L. (Gentianaceae): a comparison in three environments. Evolution 44: 1129–1139.

Dudash, M. R., D. E. Carr, and C. B. Fenster. 1997. Five generations of enforced selfing and outcrossing in *Mimulus guttatus*: inbreeding depression variation at the population and family level. Evolution 51: 54–65.

East, E. M. 1916. Studies on size inheritance in *Nicotiana*. Genetics 1: 164–176.

Eckenwalder, J. E. 1984. Natural intersectional hybridization between North American species of *Populus* (Salicaceae) in section *Aigeiros* and *Tacamahaca*. III. Paleobotany and evolution. Canad. J. Bot. 62: 336–342.

Eckert, C., and S. C. H. Barrett. 1994. Inbreeding depression in partially self-fertilizing *Decodon verticillatus* (Lythraceae): population-genetic and experimental analyses. Evolution 48: 952–964.

Ehrendorfer, F. 1968. Geographical and ecological aspects of intraspecific differentiation. Pages 261–296 in V. H. Heywood, ed., Modern Methods in Plant Taxonomy. Academic Press, London.

Ehrendorfer, F. 1980. Polyploidy and distribution. Pages 45–59 in W. H. Lewis, ed., Polyploidy—Biological Relevance. Plenum, New York.

Ehrendorfer, F. 1984. Artbegriff und Artbildung in botanischer Sicht. Z. Zool. Syst. Evolut.-forsch. 22: 234–263.

Ehrlich, P. R., and P. H. Raven. 1969. Differentiation of populations. Science 165: 1228–1232.

El Mousadik, A., and R. J. Petit. 1996. Chloroplast DNA phylogeography of the argan tree of Morocco. Molec. Ecol. 5: 547–555.

Eldredge, N. 1979. Alternative approaches to evolutionary theory. Bull. Carnegie Mus. Nat. Hist. 13: 7–19.

Eldredge, N. 1989. Macroevolutionary Dynamics. McGraw Hill, New York.

Eldredge, N. 1995. Reinventing Darwin. John Wiley & Sons, New York.

Eldredge, N., and J. Cracraft. 1980. Phylogenetic Patterns and the Evolutionary Process: Methods and Theory in Comparative Biology. Columbia University Press, New York.

Eldredge, N., and S. J. Gould. 1972. Punctuated equilibria: an alternative to phyletic gradualism. Pages 82–115 in T. J. M. Schoph, ed., Models in Paleobiology. Freeman Cooper, San Francisco.

Ellstrand, N. C. 1992a. Gene flow among seed plant populations. New Forests 6: 241–256.

Ellstrand, N. C. 1992b. Gene flow by pollen: implications for plant conservation genetics. Oikos 63: 77–86.

Ellstrand, N. C., and D. R. Elam. 1993. Population genetic consequences of small population size: implications for plant conservation. Ann. Rev. Ecol. Syst. 24: 217–242.

Ellstrand, N. C., and D. L. Marshall. 1985. Interpopulation gene flow by pollen in wild radish, *Raphanus sativus*. Amer. Natur. 126: 606–616.

Elton, C. S. 1958. The Ecology of Invasion by Animals and Plants. Chapman & Hall, London.

Emerson, S. 1939. A preliminary survey of the *Oenothera organensis* population. Genetics 24: 538–552.

Emerson, S. 1940. Growth of incompatible pollen tubes in *Oenothera organensis*. Bot. Gaz. 101: 890–911.

Endler, J. A. 1977. Geographic Variation, Speciation, and Clines. Princeton University Press, Princeton.

Endler, J. A. 1986. Natural Selection in the Wild. Princeton University Press, Princeton, New Jersey.

Ennos, R. A. 1994. Estimating the relative rates of pollen and seed migration among plant populations. Heredity 72: 250–259.

Epling, C., H. Lewis, and F. M. Ball. 1960. The breeding group and seed storage: a study in population dynamics. Evolution 14: 238–255.

Ericson, L., J. J. Burdon, and A. Wennstrom. 1993. Inter-specific host hybrids and phalacrid beetles implicated in the local survival of smut pathogens. Oikos 68: 393–400.

Eriksson, O. 1996. Regional dynamics of plants: a review of evidence for remnant, source-sink, and metapopulations. Oikos 77: 248–258.

Ernst, W. H. O. 1987. Scarcity of flower colour polymorphism in field populations of *Digitalis purpurea* L. Flora 179: 231–239.

Esselman, E. J., and D. J. Crawford. 1997. Molecular and morphological evidence for the origin of *Solidago albopilosa* (Asteraceae), a rare endemic of Kentucky. Syst. Bot. 22: 245–257.

Ewens, W. J. 1990. The minimum viable population size as a genetic and demographic concept. Pages 307–316 in J. Adams, D. A. Lamm, A. I. Hermalin, and P. E. Smouse, eds., Convergent Issues in Genetics and Demography. Oxford University Press, New York.

Falconer, D. S. 1981. Introduction to Quantitative Genetics. 2nd ed. Longman, New York.

Farrell, B. D., D. E. Dussourd, and C. Mitter. 1991. Escalation of plant defense: do latex and resin canals spur plant diversification? Amer. Natur. 138: 881–900.

Fastie, C. L. 1995. Causes and ecosystem consequences of multiple pathways of primary succession at Glacier Bay, Alaska. Ecology 76: 1899–1916.

Feinsinger, P. 1983. Coevolution and pollination. Pages 282–310 in D. Futuyma and M. Slatkin, eds., Coevolution. Sinauer Associates, Sunderland, Mass.

Felber, F. 1991. Establishment of a tetraploid cytotype in a diploid population: effect of the relative fitness of the cytotypes. J. Evol. Biol. 4: 195–207.

Fenster, C. B., and D. E. Carr. 1997. Genetics of sex allocation in *Mimulus* (Scrophulariaceae). J. Evol. Biol. 10: 641–661.

Fenster, C. B., P. K. Dingle, S. C. H. Barrett, and K. Ritland. 1995. The genetics of floral development differentiating two species of *Mimulus* (Scrophulariaceae). Heredity 74: 258–266.

Fenster, C. B., and M. R. Dudash. 1995. Genetic considerations for plant population restoration and conservation. Pages 34–62 in M. L. Bowles and C. J. Whelan, eds., Restoration of Endangered Species. Cambridge University Press, London.

Fenster, C. B., and K. Ritland. 1992. Chloroplast DNA and isozyme diversity in two *Mimulus* species (Scrophulariaceae) with contrasting mating systems. Amer. J. Bot. 79: 1440–1447.

Fiedler, P. L., and L. J. Ahouse. 1992. Hierarchies of cause: toward an understanding of rarity in vascular plant species. Pages 24–47 in P. L. Fiedler and S. K. Jain, eds., Conservation Biology. Chapman & Hall, New York.

Fischer, M., and D. Matthies. 1997. Mating structure and inbreeding and outbreeding depression in the rare plant *Gentianella germanica* (Gentianaceae). Amer. J. Bot. 84: 1685–1692.

Fischer, M., and D. Matthies. 1998a. Effects of population size on performance in the rare plant *Gentianella germanica*. J. Ecol. 86: 195–204.

Fischer, M., and D. Matthies. 1998b. RAPD variation in relation to population size and plant fitness in the rare *Gentianella germanica* (Gentianaceae). Amer. J. Bot. 85: 811–819.

Fisher, R. A. 1937. The wave of advance of advantageous genes. Ann. Eugenics 7: 355–369.

Fisher, R. A. 1958. The Genetical Theory of Natural Selection. 2nd ed. Dover, New York.

Fix, A. G. 1978. The role of kin-structured migration in genetic microdifferentiation. Ann. Human Genet. 41: 329–339.

Floate, K. D., and T. G. Whitham. 1993. The "hybrid bridge" hypothesis: host shifting via plant hybrid swarms. Amer. Natur. 141: 651–662.

Forcella, F., J. T. Wood, and S. P. Dillon. 1986. Characteristics distinguishing invasive weeds within *Echium*. Weed Res. 26: 351–364.

Ford, V. S., and L. D. Gottlieb. 1990. Genetic studies of floral evolution in *Layia*. Heredity 64: 29–44.

Fowler, C. W., and J. A. MacMahon. 1982. Selective extinction and speciation: their influence on the structure and functioning of communities and ecosystems. Amer. Natur. 119: 480–498.

Fowler, N. L., and D. A. Levin. 1984. Ecological constraints on the establishment of a novel polyploid in competition with its diploid ancestor. Amer. Natur. 124: 703–711.

Francisco-Ortega, J., R. K. Jansen, and A. Santos-Guerra. 1996a. Chloroplast DNA evidence of colonization, adaptive radiation, and hybridization in the evolution of the Macaronesian flora. Proc. Natl. Acad. Sci. USA. 93: 4085–4090.

Francisco-Ortega, J., D. J. Crawford, A. Santos-Guerra, and J. A. Carvalho. 1996b. Isozyme differentiation in the endemic genus *Argyranthemum* (Asteraceae: Anthemideae) in the Macaronesian Islands. Plant Syst. Evol. 202: 137–152.

Francisco-Ortega, J., D. J. Crawford, A. Santos-Guerra, and R. K. Jansen. 1997. Origin and evolution of *Argyranthemum* (Asteraceae: Anthemideae) in Macaronesia. Pages 407–431 in T. J. Givnish and K. J. Sytsma, eds., Molecular Evolution and Adaptive Radiation. Cambridge University Press, London.

Frankel, R. 1983. Heterosis: Reappraisal of Theory and Practice. Springer-Verlag, New York.

Friar, E. A., R. H. Robichaux, and D. W. Mount. 1996. Molecular genetic variation following a population crash in the endangered Mauna Kea silversword, *Argyroxiphium sandwicense* ssp. *sandwicense* (Asteraceae). Molec. Ecol. 5: 687–691.

Friedman, S. T., and W. T. Adams. 1985. Estimation of gene flow into two seed orchards of loblolly pine (*Pinus taeda*). Theor. Appl. Genet. 69: 609–615.

Fritsch, P. 1996. Isozyme analysis of intercontinental disjuncts within *Styrax* (Styracaceae): implications for the Madrean-Tethyan hypothesis. Amer. J. Bot. 83: 342–355.

Fritz, J. H. 1998. Artificial selection on interspecific crossing barriers in *Phlox*. Ph.D. Thesis. University of Texas, Austin.

Fritz, R. S. 1999. Resistance of hybrid plants to herbivores: genes, environment, or both? Ecology 80: 382–391.

Fritz, R. S., C. M. Nichols-Orians, and S. J. Bronsfeld. 1994. Interspecific hybridization of plants and resistance to herbivores: hypotheses, genetics, and variable responses in a diverse herbivore community. Oecologia 97: 106–117.

Fukuoka, S., H. Namai, and K. Okuno. 1998. RFLP mapping of the gene controlling hybrid breakdown in rice (*Oryza sativa* L.). Theor. Appl. Genet. 97: 446–449.

Fuller, J. L. 1998. Ecological impact of the mid-Holocene hemlock decline in southern Ontario, Canada. Ecology 79: 2337–2351.

Funk, V. A. 1982. Systematics of *Montanoa* (Asteraceae: Heliantheae). Mem. New York Bot. Gard. 36: 1–13

Funk, V. A., and D. R. Brooks. 1990. Phylogenetic systematics as the basis for comparative biology. Smithsonian Contributions in Botany 73: 1–45.

Futuyma, D. J. 1986. Evolutionary Biology. 2nd ed. Sinauer Associates, Sunderland, Mass.

Futuyma, D. J. 1987. On the role of species in anagenesis. Amer. Natur. 130: 465–473.

Gajewski, W. 1950. The inheritance of species traits in the hybrid *Geum coccineum* × *rivale*. Acta Soc. Bot. Poloniae 20: 455–476.

Galen, C. 1995. The evolution of floral form: insights from an alpine wildflower, *Polemonium viscosum* (Polemoniaceae). Pages 273–291 in D. G. Lloyd and S. C. H. Barrett, eds., Floral Biology. Chapman & Hall, New York.

Galen, C. 1996. Rates of floral evolution: adaptation to bumblebee pollination in an alpine flower, *Polemonium viscosum*. Evolution 50: 120–125.

Galen, C., and T. Gregory. 1989. Interspecific pollen transfer as a mechanism of competition: consequences of foreign pollen contamination for seed set in the alpine wildflower, *Polemonium viscosum*. Oecologia 81: 120–123.

Ganders, F. R., and K. M. Nagata. 1984. The role of hybridization in the evolution of *Bidens* (Asteraceae) on the Hawaiian Islands. Pages 179–184 in W. F. Grant, ed., Plant Biosystematics. Academic Press, Orlando.

Gaston, K. J. 1990. Patterns in the geographical ranges of species. Biol. Rev. 65: 105–129.

Gaston, K. J. 1994. Rarity. Chapman & Hall, New York.

Gaston, K. J., and W. E. Kunin. 1997. Rare-common differences: an overview. Pages 12–29 in W. E. Kunin and K. J. Gaston, eds., The Biology of Rarity. Chapman & Hall, London.

Geber, M. A. 1990. The cost of meristem limitation in *Polygonum arenastrum*: negative genetic correlations between fecundity and growth. Evolution 44: 799–819.

Geber, M. A., and T. E. Dawson. 1993. Evolutionary responses of plants to global change. Pages 179–197 in P. M. Kareiva, J. G. Kingsolver, and R. B. Huey, eds., Biotic Interactions and Global Change. Sinauer Associates, Sunderland, Mass.

Gemmill, C. E. C., T. A. Ranker, D. Ragone, S. P. Perlman, and K. R. Wood. 1998. Conservation genetics of the endangered Hawaiian genus *Brighamia* (Campanulaceae). Amer. J. Bot. 85: 528–539.

Gentry, A. H. 1986. Endemism in tropical versus temperate plant communities. Pages 153–181 in M. E. Soulé, ed., Conservation Biology: The Science of Scarcity and Diversity. Sinauer, Sunderland, Mass.

Gentry, A. H. 1990. Evolutionary patterns in the neotropical Bignoniaceae. Mem. New York Bot. Gard. 55: 118–129.

Gerstel, D. U. 1954. A new lethal combination in interspecific cotton hybrids. Genetics 39: 628–639.

Giles, B. E., and J. Goudet. 1997. A case study of genetic structure in a plant metapopulation. Pages 429–454 in I. A. Hanski and M. E. Gilpin, eds., Metapopulation Biology. Academic Press, New York.

Gili, C., and J. Martinell. 1994. Relationship between species longevity and larval ecology in nasariid gastropods. Lethaia 27: 291–299.

Gilissen, L. J., and H. F. Linskens. 1975. Pollen tube growth in styles of self-incompatible *Petunia* pollinated with radiated pollen and foreign pollen mixtures. Pages 201–205 in D. L. Mulcahy, ed., Gamete Competition in Plants and Animals. North Holland, Amsterdam.

Gillett, G. W. 1972. The role of hybridization in the evolution of the Hawaiian flora. Pages 205–219 in D. H. Valentine, eds., Taxonomy, Phytogeography, and Evolution. Academic Press, London.

Gillett, G. W., and E. K. S. Kim. 1970. An experimental study of the genus *Bidens* (Asteraceae) in the Hawaiian Islands. University California Publ. Bot. 56: 1–63.

Gilmer, K., and J. W. Kadereit. 1989. The biology and affinities of *Senecio teneriffae* Schultz Bip., an annual endemic from the Canary Islands. Bot. Jahr. 11: 263–273.

Gilpin, M. E. 1990. Extinction of finite metapopulations in correlated environments. Pages 177–186 in B. Shorrocks and I. R. Swingland, eds., Living in a Patchy Environment. Oxford University Press, New York.

Gilpin, M. E., and M. E. Soulé. 1986. Minimum viable populations: processes of species extinction. Pages 19–34 in M. E. Soulé, ed., Conservation Biology: The Science of Scarcity and Diversity. Sinauer Associates, Sunderland, Mass.

Givnish, T. J. 1998. Adaptive plant evolution on islands: classical patterns, molecular data, new insights. Pages 281–304 in P. R. Grant, ed., Evolution on Islands. Oxford University Press, New York.

Givnish, T. J., K. J. Sytsma, J. F. Smith, and W. J. Hahn. 1995. Molecular evolution, adaptive radiation, and geographic speciation in *Cyanea* (Campanulaceae, Lobelioideae). Pages 288–337 in W. L. Wagner and V. A. Funk, eds., Hawaiian Biogeography—Evolution in a Hot Spot Archipelago. Smithsonian Institution Press, Washington, D.C.

Glazier, D. S. 1987. Toward a predictive theory of speciation: the ecology of isolate selection. J. Theor. Biol. 126: 323–333.

Gleba, Y. Y., and K. M. Sytnik. 1984. Protoplast Fusion. Springer-Verlag, New York.

Glover, D. E., and S. C. H. Barrett. 1987. Genetic variation in continental and island populations of *Eichhornia paniculata*. Heredity 59: 7–17.

Godt, M. J. W., and J. L. Hamrick. 1993. Patterns and levels of pollen-mediated gene flow in *Lathyrus latifolius*. Evolution 47: 98–110.

Godt, M. J. W., J. L. Hamrick, and S. Bratton. 1995. Genetic diversity in a threatened wetland species, *Helonias bullata* (Liliaceae). Conserv. Biol. 9: 596–604.

Goldblatt, P., J. C. Manning, and P. Bernhardt. 1995. Pollination of *Lapeirousia* subgenus *Lapeirousia* (Iridaceae) in southern Africa: floral divergence and adaptation for long-tongued fly pollination. Ann. Missouri Bot. Gard. 82: 517–534.

Goldblatt, P., and J. C. Manning. 1996. Phylogeny and speciation in *Lapeirousia* subgenus *Lapeirousia* (Iridaceae:Ixioideae). Ann. Missouri Bot. Gard. 83: 346–361.

Goodell, K., D. R. Elam, J. D. Nason, and N. C. Ellstrand. 1997. Gene flow among small populations of a self-incompatible plant: an interaction between demography and genetics. Amer. J. Bot. 84: 1362–1371.

Goodman, D. 1987. The demography of chance extinction. Pages 11–34 in M. E. Soulé, ed., Viable Populations for Conservation. Cambridge University Press, New York.

Gottlieb, L. D. 1973. Genetic differentiation, sympatric speciation and the origin of a diploid species of *Stephanomeria*. Amer. J. Bot. 60: 545–553.

Gottlieb, L. D. 1974. Genetic confirmation of the origin of *Clarkia lingulata*. Evolution 28: 244–250.

Gottlieb, L. D. 1984. Genetics and morphological evolution in plants. Amer. Natur. 123: 681–709.

Gottlieb, L. D., and J. P. Bennett. 1983. Interference between individuals in pure and mixed cultures of *Stephanomeria malheurensis* and its progenitor. Amer. J. Bot. 70: 276–284.

Gottlieb, L. D., and G. Pilz. 1976. Genetic similarity between *Gaura longiflora* and its obligately outcrossing derivative *G. demareei*. Syst. Bot. 1: 181–187.

Gottlieb, L. D., S. I. Warwick, and V. S. Ford. 1985. Morphological and electrophoretic divergence between *Layia discoidea* and *L. glandulosa*. Syst. Bot. 10: 484–495.

Gould, S. J. 1980. Is a new general theory of evolution emerging? Paleobiology 6: 119–130.

Grant, P. R. 1972. Convergent and divergent character displacement. Biol. J. Linn. Soc. 4: 39–68.

Grant, V. 1952. Genetic and taxonomic studies in *Gilia*. II. *Gilia capitata abrotanifolia*. Aliso 2: 361–373.

Grant, V. 1954. Genetic and taxonomic studies in *Gilia* IV. *Gilia achilleaefolia*. Aliso 3: 1–18.

Grant, V. 1956. The genetic structure of species and races. Adv. Genet. 8: 55–87.

Grant, V. 1964. The biological composition of a taxonomic species. Adv. Genet. 12: 281–328.

Grant, V. 1966*a*. The selective origin of incompatibility barriers in the plant genus *Gilia*. Amer. Natur. 100: 99–118.

Grant, V. 1966*b*. The origin of a new species of *Gilia* in a hybridization experiment. Genetics 54: 1189–1199.

Grant, V. 1981. Plant Speciation. 2nd ed. Columbia University Press, New York.

Grant, V., and K. A. Grant, 1965. Flower Pollination in the Phlox Family. Columbia University Press, New York.

Grant, V., and D. H. Wilken. 1987. Secondary intergradation between *Ipomopsis aggregata candida* and *I. a. aggregata* (Polemoniaceae) in Colorado. Bot. Gaz. 148: 372–378.

Gray, A. J. 1986. Do invading species have definable genetic characteristics? Phil. Trans. Royal Soc. Lond. B. 314: 655–674.

Gray, A. J., D. F. Marshall, and A. F. Raybould. 1991. A century of evolution in *Spartina anglica*. Adv. Ecol. Res. 21: 1–62.

Green, J. M., and M. D. Jones. 1953. Isolation of cotton for seed increase. Agron. J. 45: 366–368.

Greuter, W. 1995. Extinctions in the Mediterranean areas. Pages 88–97 in J. H. Lawton and R. M. May, eds., Extinction Rates. Oxford University Press, New York.

Groom, M. J. 1998. Allee effects limit population viability of an annual plant. Amer. Natur. 151: 487–496.

Grun, P. 1961. Early stages in the formation of internal barriers to gene exchange between diploid species of *Solanum*. Amer. J. Bot. 48: 79–89.

Guerrant, E. R., Jr. 1992. Genetic and demographic considerations in the sampling and reintroduction of rare species. Pages 322–344 in P. L. Fiedler and S. K. Jain, eds., Conservation Biology. Chapman & Hall, New York.

Guries, R. P., and F. T. Ledig. 1982. Genetic diversity and population structure in pitch pine (*Pinus rigida*). Evolution 36: 387–402.

Gustafsson, M. 1974. Evolutionary trends in the *Atriplex triangularis* group. III. Effects of population size and introgression on chromosomal differentiation. Bot. Notiser 127: 125–148.

Guttman, S. I., and L. A. Weigt. 1989. Electrophoretic evidence of relationships among *Quercus* (oaks) of eastern North America. Canad. J. Bot. 67: 339–351.

Haase, P. 1993. Genetic variation, gene flow, and the "founder effect" in pioneer populations of *Nothofagus menziesii* (Fagaceae), South Island, New Zealand. J. Biogeogr. 20: 79–85.

Hamilton, C. W. 1990. Variations on a distylous theme in Mesoamerican *Psychotria* subgenus *Psychotria* (Rubiaceae). Memoirs New York Bot. Gard. 55: 62–75.

Hamrick, J. L. 1989. Isozymes and analysis of genetic structure of plant populations. Pages 87–105 in D. Soltis and P. Soltis, eds., Isozymes in Plant Biology. Dioscorides Press, Washington, D.C.

Hamrick, J. L., H. M. Blanton, and K. J. Hamrick. 1989. Genetic structure of geographically marginal populations of ponderosa pine. Amer. J. Bot. 76: 1559–1568.

Hamrick, J. L., and M. J. W. Godt. 1990. Allozyme diversity in plant species. Pages 43–63 in A. H. D. Brown, M. T. Clegg, A. L. Kahler, and B. S. Weir, eds., Plant Population Genetics, Breeding, and Genetic Resources. Sinauer Associates, Sunderland, Mass.

Hamrick J. L., and M. W. Godt. 1995. Conservation genetics of endemic plant species. Pages 281–304 in J. C. Avise and J. L. Hamrick, eds., Conservation Genetics: Case Histories from Nature. Chapman & Hall, New York.

Hamrick, J. L., and M. J. W. Godt. 1996. Effects of life history traits on genetic diversity in plant species. Phil. Trans. Royal. Soc. Lond. B. 351: 1291–1298.

Hamrick, J. L., M. J. W. Godt, and S. Sherman-Broyles. 1995. Gene flow among plant populations: evidence from genetic markers. Pages 215–232 in P. C. Hoch and A. G. Stephenson, eds., Experimental and Molecular Approaches to Plant Systematics. Missouri Botanical Garden, St. Louis.

Hamrick, J. L., and M. D. Loveless. 1989. The genetic structure of tropical tree populations: associations with reproductive biology. Pages 129–146 in J. H. Bock and Y. B. Linhart, eds., The Evolutionary Ecology of Plants. Westview Press, Boulder, Colo.

Handel, S. N. 1982. Dynamics of gene flow in an experimental population of *Cucumis melo* (Cucurbitaceae). Amer. J. Bot. 69: 1538–1546.

Handel, S. N. 1983. Contrasting gene flow patterns and genetic subdivision in adjacent populations of *Cucumus sativus* (Cucurbitaceae). Evolution 37: 760–771.

Hanski, I. 1989. Metapopulation dynamics: does it help to have more of the same? Trends Ecol. Evol. 4: 113–114.

Hanski, I. 1991. Single-species metapopulation dynamics: concepts, models and observations. Biol. J. Linn. Soc. 42: 17–38.

Harder, L. D., M. B. Cruzan, and J. D. Thomson. 1993. Unilateral incompatibility and the effects of interspecific pollination for *Erythronium americanum* and *Erythronium albidum* (Liliaceae). Canad. J. Bot. 71: 353–358.

Harlan, J. R. 1976. Diseases as a factor in evolution. Ann. Rev. Phytopathol. 14: 31–51.

Harper, J. L., J. N. Clatworthy, I. H. McNaughton, and G. R. Sager. 1961. The evolution and ecology of species living in the same area. Evolution 15: 209–227.

Harrison, S. 1991. Local extinction in a metapopulation context: an empirical evaluation. Biol. J. Linn. Soc. 42: 73–88.

Harrison, S., and J. F. Quinn. 1989. Correlated environments and the persistence of populations. Oikos 56: 293–298.

Hedrick, P. W., and L. Holden. 1979. Hitchhiking: an alternative to coadaptation for the barley and slender wild oat examples. Heredity 43: 79–86.

Heiser, C. B. 1961. Morphological and cytological variation in *Helianthus petiolaris* with notes on related species. Evolution 15: 247–258.

Heiser, C. B. 1979. Hybrid populations of *Helianthus divaricatus* and *H. microcephalus* after 22 years. Taxon 28: 71–75.

Helenurm, K., and F. R. Ganders. 1985. Adaptive radiation and genetic differentiation in Hawaiian *Bidens*. Evolution 39: 753–765.

Hengeveld, R. 1990. Dynamic Biogeography. Cambridge University Press, New York.

Herrera, C. M. 1989. Frugivory and seed dispersal by carnivorous mammals, and associated fruit characteristics, in undisturbed Mediterranean habitats. Oikos 55: 250–262.

Herrera, C. M. 1996. Plant traits and plant adaptation to insect pollinators: a devil's advocate approach. Pages 65–87 in D. G. Lloyd and S. C. H. Barrett, eds., Floral Biology. Chapman & Hall, New York.

Heschel, M. S., and K. N. Paige. 1995. Inbreeding depression, environmental stress, and population size variation in scarlet gilia (*Ipomopsis aggregata*). Conserv. Biol. 9: 126–133.

Heslop-Harrison, J. 1964. Forty years of genecology. Adv. Ecol. Res. 2:159–247.

Hewitt, G. M. 1988. Hybrid zones–natural laboratories for evolutionary studies. Trends Ecol. Evol. 3: 158–167.

Hewitt, G. M. 1989. The subdivision of species by hybrid zones. Pages 85–110 in D. Otte and J. A. Endler, eds., Speciation and Its Consequences. Sinauer Associates, Sunderland, Mass.

Hewitt, G. M. 1993. Postglacial distribution and species substructure: lessons from pollen, insects and hybrid zones. Pages 98–123 in Evolutionary Patterns and Processes, The Linnean Society of London.

Hewitt, G. M. 1996. Some genetic consequences of ice ages, and their role in divergence and speciation. Biol. J. Linn. Soc. 58: 247–276.

Heywood, J. S. 1986*a*. Clinal variation associated with edaphic ecotones in hybrid populations of *Gaillardia pulchella*. Evolution 40: 1132–1140.

Heywood, J. 1986*b*. The effect of plant size variation on genetic drift in annuals. Amer. Natur. 127: 851–861.

Heywood, J. S., and D. A. Levin. 1984. Allozyme variation in *Gaillardia pulchella* and *G. amblyodon* (Compositae): relation to morphological and chromosomal variation and to geographical isolation. Syst. Bot. 9: 448–457.

Heywood, V. 1979. The future of island floras. Pages 431–441 in D. Bramwell, eds., Plants and Islands. Academic Press, New York.

Hiesey, W. M., M. A. Nobs, and O. Bjorkman. 1971. Experimental studies on the nature of species. V. Biosystematics, genetics, and physiological ecology of the *Euryanthe* section of *Mimulus*. Carnegie Institution of Washington Publ. No. 628.

Hobbs, R. J. 1989. The nature and effect of disturbance relative to invasion. Pages 389–405 in J. A. Drake, H. A. Mooney, F. diCastri, R. H. Groves, F. J. Kruger, M. Rejmánek and M. Williamson, eds. Biological Invasions—A Global Perspective. John Wiley & Sons, Chichester.

Hodges, S. A. 1997. Rapid radiation due to a key innovation in columbines (Ranunculaceae: *Aquilegia*). Pages 391–405 in T. J. Givnish and K. J. Sytsma, eds., Molecular Evolution and Adaptive Radiation, Cambridge University Press, New York.

Hodges, S. A., and M. L. Arnold. 1994. Columbines: a geographically widespread species flock. Proc. Natl. Acad. Sci. USA. 91: 5129–5132.

Hoey, M. T., and C. R. Parks. 1991. Isozyme divergence between eastern Asian, North American, and Turkish species of *Liquidambar* (Hamamelidaceae). Amer. J. Bot. 78: 938–947.

Hoffmann, A. A., and M. W. Blows. 1994. Species borders: ecological and evolutionary perspectives. Trends Ecol. Evol. 9: 223–227.

Hoffmann, A. A., and P. A. Parsons. 1997. Extreme Environmental Change and Evolution. Cambridge University Press, New York.

Holdgate, M. W. 1986. Summary and conclusions: characteristics and consequences of biological invasions. Phil. Trans. Royal Soc. Lond. B: 314: 733–742.

Hollingshead, L. 1930. A lethal factor in *Crepis* effective only in interspecific hybrids. Genetics 15: 114–140.

Holsinger, K. E. 1993. The evolutionary dynamics of fragmented plant populations. Pages 198–216 in P. M. Kareiva, J. G. Kingsolver, and R. B. Huey, eds., Biotic Interactions and Global Change. Sinauer Associates, Sunderland, Mass.

Holt, R. D., and R. Gomulkiewicz. 1997. How does immigration influence local adaptation? A reexamination of a familiar paradigm. Amer. Natur. 149: 563–572.

Holtsford, T. P. 1996. Variation in inbreeding depression among families and populations of *Clarkia tembloriensis*. Heredity 76: 83–91.

Holtsford, T. P., and N. C. Ellstrand. 1990. Inbreeding depression in *Clarkia tembloriensis* (Onagraceae) populations with different natural outcrossing rates. Evolution 44: 2031–2046.

Hopper, S. D. 1978. An experimental study of competitive interference between *Angiozanthos manglesii* D. Don, *A. humilis* Lindl. and their F_1 hybrids (Haemodoraceae). Austral. J. Bot. 26: 807–817.

Howard, D. J. 1993. Reinforcement: origin, dynamics, and fate of an evolutionary hypothesis. Pages 46–69 in R. G. Harrison, ed., Hybrid Zones and the Evolutionary Process, Oxford University Press, New York.

Howard, D. J., and S. H. Berlocher. 1998. Endless Forms: Species and Speciation. Oxford University Press, New York.

Howard, D. J., R. W. Preszler, J. Williams, S. Fenchel, and W. J. Boecklin. 1997. How discrete are oak species? Insights from a hybrid zone between *Quercus grisea* and *Quercus gambelii*. Evolution 51: 747–755.

Huenneke, L. F. 1991. Ecological implications of genetic variation in plant populations. Pages 31–44 in D. A. Falk and K. E. Holsinger, eds. Genetics and Conservation of Rare Plants. Oxford University Press, New York.

Huether, C. A. 1968. Exposure of natural genetic variability underlying the pentamerous corolla constancy in *Linanthus androsaceus* var. *androsaceus*. Genetics 60: 123–146.

Humphries, C. J. 1976a. Evolution and endemism in *Argyranthemum* Webb ex Schultz Bip. (Compositae-Anthemidae). Botanica Macaronesia 1: 25–50.

Humphries, C. J. 1976b. A revision of the Macaronesian genus *Argyranthemum* Webb ex Schultz Bip. (Compositae-Anthemidae). Bull. British Museum (Natural History) Botany 5: 147–240.

Huntley, B. 1991. How plants respond to climatic change: migration rates, individualism and the consequences for plant communities. Ann. Bot. 67 (Suppl. 1): 15s–22s.

Huntley, B., P. J. Bartlein, and I. C. Prentice. 1989. Climatic control of the distribution and abundance of beech (*Fagus* L.) in Europe and North America. J. Biogeogr. 16: 551–560.

Huntley, B., and H. J. B. Birks. 1983. An Atlas of Past and Present Pollen Maps for Europe: 0–13,000 Years Ago. Cambridge University Press, Cambridge.

Huntley, B., and T. Webb, III. 1989. Migration: species' response to climatic variations caused by changes in the earth's orbit. J. Biogeogr. 16: 5–19.

Husband, B. C., and D. W. Schemske. 1996. Evolution of the magnitude and timing of inbreeding depression in plants. Evolution 50: 54–70.

Husband, B. C., and D. W. Schemske. 1997. The effect of inbreeding in diploid and tetraploid populations of *Epilobium angustifolium* (Onagraceae): implications for the genetic basis of inbreeding depression. Evolution 51: 737–746.

Husband, B. C., and D. W. Schemske. 1998. Cytotype distribution at a diploid-tetraploid contact zone in *Chamaenerion* (*Epilobium*) *angustifolium* (Onagraceae). Amer. J. Bot. 85: 1688–1694.

Hutchinson, G. E. 1965. The Ecological Theater and the Evolutionary Play. Yale University Press, New York.

Huxley, J. S. 1940. The New Systematics. Clarendon Press, Oxford.

Imbrie, B. C., and P. F. Knowles. 1971. Genetic studies of self-incompatibility in *Carthamus flavescens* Spreng. Crop Sci. 11: 6–9.

Imbrie, B. C., J. C. Kirkman, and D. R. Ross. 1972. Computer simulations of a sporophytic incompatibility system. Austral. J. Biol. Sci. 25: 343–349.

Inoue, K., and T. Kawahara. 1990. Allozyme differentiation and genetic structure in island and mainland Japanese populations of *Campanula punctata* (Campanulaceae). Amer. J. Bot. 77: 1440–1448.

Jablonski, D. 1987. Heritability at the species level: analysis of geographical variation of cretaceous mollusks. Science 238: 360–363.

Jackson, R. C., and C. F. Dimas. 1981. Experimental evidence for the systematic placement of the *Haplopappus phyllocephalus* complex (Compositae). Syst. Bot. 6: 8–14.

Jain, S. K. 1976. The evolution of inbreeding in plants. Ann. Rev. Ecol. Syst. 7: 469–495.

Jain, S. K. 1977. Response to mass selection for flowering time in meadowfoam. Crop Sci. 19: 337–339.

Jalani, B. S., and J. P. Moss. 1980. The site of action of the crossability genes (Kr_1, Kr_2) between *Triticum* and *Secale*. I. Pollen germination, pollen tube growth, and number of pollen tubes. Euphytica 29: 571–579.

Janzen, D. H. 1981. *Enterolobium cyclocarpum* seed passage rate and survival in horses, Costa Rican Pleistocene seed dispersal agents. Ecology 63: 593–601.

Jenkins, G., and G. Jimenez. 1995. Genetic control of synapsis and recombination in *Lolium* amphidiploids. Chromosoma 104: 164–168.

Jennersten, O. 1988. Pollination in *Dianthus deltoides* (Caryophyllaceae): effects of habitat fragmentation on visitation and seed-set. Conserv. Biol. 2: 359–366.

Jennersten, O., and S. G. Nilsson. 1993. Insect flower visitation frequency and seed production in relation to patch size of *Viscaria vulgaris* (Caryophyllaceae). Oikos 68: 283–292.

Johnson, S. D., H. P. Linder, and K. E. Steiner. 1998. Phylogeny and radiation of pollination systems in *Disa* (Orchidaceae). Amer. J. Bot. 85: 402–411.

Johnson, W. C., and T. Webb. III. 1989. The role of blue jays (*Cyanocitta cristata* L.) in the postglacial dispersal of fagaceous trees in eastern North America. J. Biogeogr. 16: 561–571.

Johnston, M. O. 1991. Natural selection on floral traits in two species of *Lobelia* with different pollinators. Evolution 45: 1468–1479.

Johnston, M. O., and D. J. Schoen. 1995. Mutation rates and dominance levels of genes affecting total fitness in two angiosperm species. Science 267: 226–229.

Johnstone, I. M. 1986. Plant invasion windows: a time-based classification of invasion potential. Biol. Rev. 61: 369–394.

Jones, M. D., and J. S. Brooks. 1950. Effectiveness of Distance and Border Rows in Preventing Outcrossing in Corn. Oklahoma Agric. Expt. Sta. Tech Bull. No. T-38.

Jones, R. N., and H. Rees. 1968. Nuclear DNA variation in *Allium*. Heredity 23: 591–605.

Juvik, J. O., and S. P. Juvik. 1992. Mullein (*Verbascum thapsus*): the spread and adaptation of a temperate weed in the montane tropics. Pages 254–270 in C. P. Stone, C. W. Smith, and J. T. Tunison, eds., Alien Plant Invasions in Native Ecosystems of Hawaii: Management and Research. University of Hawaii Cooperative National Park Resources Study Unit, Honolulu, HA, USA.

Kalton, R. R., A. G. Smit, and R. C. Leffel. 1952. Parent-inbred progeny relationships of selected orchardgrass clones. Agron. J. 44: 481–486.

Kammer, T. W., T. K. Baumiller, and W. I. Ausich. 1998. Evolutionary significance of differential species longevity in Osagean-Meramecian (Mississippian) crinoid clades. Paleobiology 24: 155–176.

Karoly, K. 1992. Pollinator limitation in the facultatively autogamous annual, *Lupinus nanus* (Leguminosae). Amer. J. Bot. 79: 49–56.

Karron, J. D. 1987. A comparison of levels of genetic polymorphism and self-compatibility in geographically restricted and widespread plant congeners. Evol. Ecol. 1: 47–58.

Karron, J. D. 1991. Patterns of genetic variation and breeding systems in rare plant species.

Pages 87–98 in D. A. Falk and K. E. Holsinger, eds., Genetics and Conservation of Rare Plants. Oxford University Press, New York.

Kaul, M. L. H. 1988. Male Sterility in Higher Plants. Springer-Verlag, New York.

Keim, P., K. N. Paige, T. G. Whitham, and K. G. Lark. 1989. Genetic analysis of an interspecific hybrid swarm of Populus: occurrence of unidirectional introgression. Genetics 123: 557–565.

Kelly, J. P. 1915. Cultivated varieties of Phlox drummondii. J. New York Botanical Garden 16: 179–191.

King, J. E. 1981. Late Quaternary vegetational history of Illinois. Ecol. Monogr. 51: 43–62.

King, R. A., and C. Ferris. 1998. Chloroplast DNA phylogeography of Alnus glutinosa (L.) Gaertn. Molec. Ecol. 7: 1151–1161.

Kirkpatrick, K. J., and H. D. Wilson. 1988. Interspecific gene flow in Cucurbita: C. texana and C. pepo. Amer. J. Bot. 75: 519–527.

Kirkpatrick, M. 1982. Quantum evolution and punctuated equilibria in continuous genetic characters. Amer. Natur. 119: 833–848.

Kirkpatrick, M., and N. H. Barton. 1997. Evolution of species' range. Amer. Natur. 150: 1–23.

Kiviniemi, K. 1996. A study of adhesive seed dispersal of three species under natural conditions. Acta Bot. Neerl. 45: 73–83.

Klips, R. A. 1995. Genetic affinity of the rare eastern Texas endemic Hibiscus dasycalyx (Malvaceae). Amer. J. Bot. 82: 1463–1472.

Knowles, R. P., and H. Baenziger. 1962. Fertility indices in cross-pollinated grasses. Canad. J. Plant Sci. 42: 460–471.

Knowles, R. P., and A. W. Ghosh. 1968. Isolation requirements for smooth bromegrass, Bromus inermis, as determined by a genetic marker. Crop Sci. 3: 371–374.

Knox, E. B., and J. D. Palmer. 1995. Chloroplast DNA variation and the recent radiation of the giant senecios (Asteraceae) on the tall mountains of eastern Africa. Proc. Natl. Acad. Sci. USA 92: 10349–10353.

Kohn, J. R., S. W. Graham, B. Morton, J. J. Doyle, and S. C. H. Barrett. 1996. Reconstruction of the evolution of reproductive characters in Pontederiaceae using phylogenetic evidence from chloroplast DNA restriction-site variation. Evolution 50: 1454–1459.

Kowarik, I. 1995. Time lags in biological invasions with regard to the success and failure of alien species. Pages 15–38 in P. Pysek, K. Prach, M. Rejmánek, and M. Wade, eds., Plant Invasions—General Aspects and Special Problems. SPB Academic Publishing, Amsterdam.

Kot, M., M. A. Lewis, and P. van den Driessche. 1996. Dispersal data and the spread of invading organisms. Ecology 77: 2027–2042.

Kruckeberg, A. R. 1957. Variation in fertility of hybrids between isolated populations of the serpentine species, Strepthanthus glandulosus Hook. Evolution 11: 185–211.

Kruckeberg, A. R. 1961. Artificial crosses of western North American silenes. Brittonia 13: 305–333.

Kruckeberg, A. R., and D. Rabinowitz. 1985. Biological aspects of endemism in higher plants. Ann. Rev. Ecol. Syst. 16: 447–479.

Kuittinen, H., M. J. Silanpää, and O. Savolainen. 1997. Genetic basis of adaptation: flowering time in Arabidopsis thaliana. Theor. Appl. Genet. 95: 573–583.

Kunin, W. E. 1997. Population biology and rarity: on the complexity of density-dependence in insect-plant interactions. Pages 150–173 in W. E. Kunin and K. J. Gaston, eds., The Biology of Rarity, Chapman & Hall, London.

Kunin, W. E., and Y. Iwasa. 1996. Pollinator foraging strategies in mixed floral arrays: density effects and flower constancy. Theor. Pop. Biol. 49: 232–263.

Kunin, W. E., and A. Shmida. 1997. Plant reproductive traits as a function of local, regional, and global abundance. Conserv. Biol. 11: 183–191.

Kyhos, D. W. 1965. The independent aneuploid origin of two species of *Chaenactis* (Compositae) from a common ancestor. Evolution 19: 26–43.

Ladizinsky, G., D. Cohen, and F. J. Muehlbauer. 1985. Hybridization in the genus *Lens* by means of embryo culture. Theor. Appl. Genet. 70: 97–101.

Lagercrantz, U., and D. J. Lyiate. 1996. Comparative genome mapping in *Brassica*. Genetics 144: 1903–1910.

Lahti, T., E. Kemppainen, A. Kutro, and P. Uotila. 1991. Distribution and biological characteristics of threatened vascular plants in Finland. Biol. Conserv. 55: 299–314.

Lambert, D. M., M. R. Centner, and H. E. H. Paterson. 1984. A simulation of the conditions necessary for the evolution of species by reinforcement. South African J. Sci. 80: 308–311.

Lammers, T. G. 1990. Campanulaceae. Pages 420–489 in W. L. Wagner, D. R. Herbst, and S. H. Sohmer, eds., Manual of the Flowering Plants of Hawaii. University of Hawaii Press and Bishop Museum Press, Honolulu.

Lammers, T. G. 1995. Patterns of speciation and biogeography in *Clermontia* (Campanulaceae, Lobelioideae). Pages 338–362 in W. L. Wagner and V. A. Funk, eds., Hawaiian Biogeography—Evolution on a Hot Spot Archipelago. Smithsonian Institution Press, Washington, D.C.

Lamont, B. B., P. G. L. Klinkhamer, and E. T. F. Witkowski. 1993. Population fragmentation may reduce fertility to zero in *Banksia goodii*—a demonstration of the Allee effect. Oecologia 94: 446–450.

Lande, R. 1979a. Effective deme sizes during long-term evolution estimated from rates of chromosomal rearrangement. Evolution 33: 463–479.

Lande, R. 1979b. Quantitative genetic analysis of multivariate evolution, applied to brain size: body size allometry. Evolution 33: 402–416.

Lande, R. 1980. Genetic variation and phenotypic evolution during allopatric speciation. Amer. Natur. 116: 463–479.

Lande, R. 1983. The response to selection on major and minor mutations affecting a metrical trait. Heredity 50: 47–65.

Lande, R. 1984. The expected fixation rate of chromosomal inversions. Evolution 38: 743–752.

Lande, R. 1985. The fixation of chromosomal rearrangements in a subdivided population with local extinction and colonization. Heredity 54: 323–332.

Lande, R. 1986. The dynamics of peak shifts and the pattern of morphological evolution. Paleobiology 12: 343–354.

Lande, R. 1988. Genetics and demography in biological conservation. Science 241: 1455–1460.

Lande, R., and D. W. Schemske. 1985. The evolution of self-fertilization and inbreeding depression in plants. I. Evolution 39: 24–40.

Langlet, O. 1971. Two hundred years of genecology. Taxon 20: 653–722.

Latta, R., and K. Ritland. 1994. The relationship between inbreeding depression and prior inbreeding among populations of four *Mimulus* taxa. Evolution 48: 806–817.

Lawton, J. H. 1995. Population dynamic principles. Pages 147–163 in J. H. Lawton and R. M. May, ed., Extinction Rates. Oxford University Press, Oxford.

Le Corre, V., N. Machon, R. J. Petit, and A. Kremer. 1997. Colonization with long-distance seed dispersal and genetic structure of maternally inherited genes in forest trees: a simulation study. Genetical Research 69: 117–125.

Ledig, F. T., and M. T. Conkle. 1983. Gene diversity and genetic structure in a narrow endemic, Torrey pine (*Pinus torreyana* Parry ex Carr). Evolution 37: 79–85.

Ledig, F. T., V. Jacob-Cervantes, P. D. Hodgekiss, and T. Eguiluz-Piedra. 1997. Recent evolution and divergence among populations of a rare Mexican endemic, Chihuahua spruce, following climatic Holocene warming. Evolution 51: 1815–1827.

Lehmann, N. L. 1987. Floral formula variation in *Phlox drummondii* Hook. Masters Thesis. University of Texas, Austin.

Leroi, A. M., M. R. Rose, and G. V. Lauder. 1994. What does the comparative method reveal about adaptation? Amer. Natur. 143: 381–402.

Les, D. H., J. A. Reinhartz, and E. J. Esselman. 1991. Genetic consequences of rarity in *Aster furcatus* (Asteraceae), a threatened, self-incompatible plant. Evolution 45: 1641–1650.

Lesica, P., and F. W. Allendorf. 1992. Are small populations worth preserving? Conserv. Biol. 6: 135–139.

Levin, D. A. 1967. Variation in *Phlox divaricata*. Evolution 21: 92–108.

Levin, D. A. 1970. Developmental instability and evolution in peripheral isolates. Amer. Natur. 104: 343–353.

Levin, D. A. 1972. Low frequency disadvantage in the exploitation of pollinators by corolla variants in *Phlox*. Amer. Natur. 106: 453–460.

Levin, D. A. 1975. Minority cytotype exclusion in local plant populations. Taxon 24: 35–43.

Levin, D. A. 1976a. Consequences of long-term artificial selection, inbreeding and isolation in *Phlox*. I. The evolution of cross-incompatibility. Evolution 30: 335–344.

Levin, D. A. 1976b. Consequences of long-term artificial selection, inbreeding and isolation in *Phlox*. II. The organization of allozymic variability. Evolution 30: 463–472.

Levin, D. A. 1977. The organization of genetic diversity in *Phlox drummondii*. Evolution 31: 477–494.

Levin, D. A. 1978a. The origin of isolating mechanisms in flowering plants. Evol. Biol. 11: 185–317.

Levin, D. A. 1978b. Pollinator behaviour and the breeding structure of plant populations. Pages 133–150 in A. J. Richards, ed., The Pollination of Flowers in Insects. Academic Press, London.

Levin, D. A. 1979. The nature of plant species. Science 204: 381–384.

Levin, D. A. 1983a. Polyploidy and novelty in flowering plants. Amer. Natur. 122: 1–25.

Levin, D. A. 1983b. An immigration-hybridization episode in *Phlox*. Evolution 37: 575–581.

Levin, D. A. 1984. Genetic variation and divergence in a disjunct *Phlox*. Evolution 38: 223–225.

Levin, D. A. 1985. Reproductive character displacement in *Phlox*. Evolution 39: 1275–1281.

Levin, D. A. 1989. Inbreeding depression in partially self-fertilizing *Phlox*. Evolution 43: 1417–1423.

Levin, D. A. 1993. Local speciation in plants: the rule not the exception. Syst. Bot. 18: 197–208.

Levin, D. A. 1995. Plant outliers: an ecogenetic perspective. Amer. Natur. 145: 109–118.

Levin, D. A., and E. T. Brack. 1995. Natural selection against white petals in *Phlox*. Evolution 49: 1017–1022.

Levin, D. A., and Z. Bulinska-Radomska. 1988. Effects of hybridization and inbreeding on fitness in *Phlox*. Amer. J. Bot. 75: 1632–1639.

Levin, D. A., and K. Clay. 1984. Dynamics of synthetic *Phlox drummondii* populations at the species margin. Amer. J. Bot. 71: 1040–1050.

Levin, D. A., J. Francisco-Ortega, and R. K. Jansen. 1996. Hybridization and the extinction of rare plant species. Conserv. Biol. 10: 10–16.

Levin, D. A., and H. W. Kerster. 1967. Natural selection for reproductive isolation in *Phlox*. Evolution 21: 679–687.

Levin, D. A., and H. W. Kerster. 1974. Gene flow in seed plants. Evol. Biol. 7: 139–220.

Levin, D. A., and H. W. Kerster. 1975. The effect of gene dispersal on the dynamics and statics of gene substitution in plants. Heredity 35: 317–336.

Levin, D. A., and M. Levy. 1971. Secondary intergradation and genome incompatibility in *Phlox pilosa* (Polemoniceae). Brittonia 23: 246–265.

Levin, D. A., K. Ritter, and N. C. Ellstrand. 1979. Protein polymorphism in the narrow endemic *Oenothera organensis*. Evolution 33: 534–542.

Levin, D. A., and B. A. Schaal. 1970. Corolla color as an inhibitor of interspecific hybridization in *Phlox*. Amer. Natur. 194: 273–283.

Levin, D. A., and K. P. Schmidt. 1985. Dynamics of a hybrid zone in *Phlox*: an experimental demographic investigation. Amer. J. Bot. 72: 1404–1409.

Levin, D. A., and A. C. Wilson. 1976. Rates of chromosomal evolution in seed plants: net increase in diversity of chromosome numbers and species numbers through time. Proc. Natl. Acad. Sci. USA. 73: 2086–2090.

Levins, R. 1970. Extinction. Pages 77–107 in M. Gerstenhaber, ed., Some Mathematical Problems in Biology. American Mathematical Society. Providence, R.I.

Levy, F. 1991. A genetic analysis of reproductive barriers in *Phacelia dubia*. Heredity 67: 331–345.

Lewis, H. 1961. Experimental sympatric populations of *Clarkia*. Amer. Natur. 95: 155–168.

Lewis, H. 1962. Catastrophic selection as a factor in speciation. Evolution 16: 257–271.

Lewis, H. 1966. Speciation in flowering plants. Science 152:167–172.

Lewis, H. 1967. The taxonomic significance of autopolyploidy. Taxon 16: 267–271.

Lewis, H. 1973. The origin of diploid neospecies in *Clarkia*. Amer. Natur. 107: 161–170.

Lewis, H., and M. E. Lewis. 1955. The genus *Clarkia*. University California Publ. Bot. 20: 241–392.

Lewis, H., and M. R. Roberts. 1956. The origin of *Clarkia lingulata*. Evolution 10: 126–138.

Lewis, P. O., and D. J. Crawford. 1995. Pleistocene refugium endemics exhibit greater allozyme diversity than widespread congeners in the genus *Polygonella* (Polygonaceae). Amer. J. Bot. 82: 141–149.

Lewis, W. H. 1980. Polyploidy in species populations. Pages 103–144 in W. H. Lewis, ed., Polyploidy–Biological Relevance. Plenum Press, New York.

Li, P., and W. T. Adams. 1989. Range-wide patterns of allozymic variation in Douglas-fir (*Pseudotsuga menziesii*). Canad. J. Forest Res. 19: 149–161.

Li, Z., S. R. M. Pinson, A. H. Paterson, W. D. Park, and J. W. Stansel. 1997. Genetics of hybrid sterility and hybrid breakdown in an intersubspecific rice (*Oryza sativa*) population. Genetics 145: 1139–1145.

Lin, Y.-R., K. F. Schertz, and A. H. Paterson. 1995. Comparative analysis of QTLs affecting plant height and maturity across the Poaceae, in reference to an interspecific sorghum population. Genetics 141: 391–411.

Lindsay, D. W., and R. K. Vickery, Jr. 1967. Comparative evolution in *Mimulus guttatus* of the Bonneville Basin. Evolution 21: 439–456.

Linhart, Y. B., and A. C. Premoli. 1994. Genetic variation in central and disjunct populations of *Lilium parryi*. Canad. J. Bot. 72: 79–85.

Liou, L. W., and T. D. Price. 1994. Speciation by reinforcement of premating isolating mechanisms. Evolution 48: 1451–1459.

Liston, A. 1992. Isozyme systematics of *Astragalus* sect. *Leptocarpi* subsect. *Californici* (Fabaceae). Syst. Bot. 17: 367–379.

Liston, A., L. H. Rieseberg, and T. S. Elias. 1989. Genetic similarity is high between intercontinental disjunct species of *Senecio* (Asteraceae). Amer. J. Bot. 76: 383–388.

Liu, B., J. M. Vega, G. Segal, S. Abbo., M. Rodova, and M. Feldman. 1998. Rapid changes in newly synthesized amphiploids of *Triticum* and *Aegilops*. I. Changes in low-copy non-coding DNA sequences. Genome 41: 272–277.

Livingstone, F. B. 1992. Gene flow in the Pleistocene. Human Biol. 64: 67–80.

Loeschcke, V., J. Tomiuk, and S. K. Jain. 1994. Introductory remarks: genetics and conservation biology. Pages 3–8 in V. Loeschcke, J. Tomiuk, and S. K. Jain, eds., Birkhäuser Verlag, Basel, Switzerland.

Lonsdale, W. M. 1993. Rates of spread of an invading species—*Mimosa pigra* in northern Australia. J. Ecol. 81: 513–521.

Lowrey, T. K. 1986. A biosystematic revision of Hawaiian *Tetramolopium* (Composite: Astereae). Allertonia 4: 203–265.

Lowrey, T. K. 1995. Phylogeny, adaptive radiation, and biogeography of Hawaiian *Tetramolopium* (Asteraceae: Astereae). Pages 195–219 in W. L. Wagner and V. A. Funk, eds., Hawaiian Biogeography-Evolution on a Hot Spot Archipelago. Smithsonian Institution Press, Washington, D.C.

Lowrey, T. K., and D. J. Crawford. 1985. Allozyme divergence and evolution in *Tetramolopium* (Compositae: Asteraceae) on the Hawaiian Islands. Syst. Bot. 10: 64–72.

Luckow, M. 1995. Species concepts: assumptions, methods, and applications. Syst. Bot. 20: 589–605.

Lumeret, R., C. M. Brown, and T. A. Dyler. 1989. Autopolyploidy in *Dactylis glomerata* L.: further evidence from studies of chloroplast DNA variation. Theor. Appl. Genet. 78: 393–399.

Lumaret, R., J.-L. Guillerm, J. Delay, A. Ait Lhaj Loutfi, J. Izco, and M. Jay. 1987. Polyploidy and habitat differentiation in *Dactylis glomerata* L. from Galicia (Spain). Oecologia 73: 436–446.

Lynch, M. 1996. A quantitative-genetic perspective on conservation issues. Pages 471–501 in J. C. Avise and J. L. Hamrick, eds., Conservation Genetics–Case Histories from Nature. Chapman & Hall, New York.

Lynch, M., J. Conery, and R. Burger. 1995. Mutational meltdowns in sexual populations. Evolution 49: 1067–1080.

MacArthur, R. H. 1972. Geographical Ecology. Harper & Row, New York.

MacArthur, R. H., and E. O. Wilson. 1967. The Theory of Island Biogeography. Princeton University Press, Princeton, N.J.

Mack, R. N. 1981. Invasion of *Bromus tectorum* L. into western North America: an ecological chronicle. Agro-ecosystems 7: 145–165.

Mack, R. N. 1985. Invading plants: their potential contribution to population biology. Pages 127–142 in J. White, ed., Studies on Plant Demography. Academic Press, London.

Macnair, M. R. 1989. The potential for rapid speciation in plants. Genome 31: 203–210.

Macnair, M. R. 1993. The genetics of metal tolerance in vascular plants. New Phytol. 124: 541–559.

Macnair, M. R., and P. Christie. 1983. Reproductive isolation as a pleiotropic effect of copper tolerance in *Mimulus guttatus*? Heredity 50: 295–302.

Macnair, M. R., and Q. J. Cumbes. 1989. The genetic architecture of interspecific variation in *Mimulus*. Genetics 122: 211–222.

Macnair, M. R., V. E. Macnair, and B. E. Martin. 1989. Adaptive speciation in *Mimulus*: an ecological comparison of *M. cupriphilus* with its presumed progenitor, *M. guttatus*. New Phytol. 112: 269–279.

Mallet, J. 1995. A species definition for the Modern Synthesis. Trends. Ecol. Evol. 10: 294–299.

Malpas, J. 1991. Serpentine and the geology of serpentinized rocks. Pages 7–30 in B. A. Roberts and J. Proctor, eds., The Ecology of Areas with Serpentinized Rocks: A World View. Kluwer, Dortrecht.

Manasse, R. 1992. Ecological risks of transgenic plants: effects of spatial dispersion on gene flow. Ecol. Applic. 2: 431–438.

Margules, C. R., and M. P. Austin. 1995. Biological models for monitoring species decline: the construction and use of data bases. Pages 183–196 in J. H. Lawton and R. M. May, eds., Extinction Rates. Oxford University Press, Oxford.

Martin, M. M., and J. Harding. 1981. Evidence for the evolution of competition between two species of annual plants. Evolution 35: 975–987.

Malo, J. E., and F. Suárez. 1995. Herbivorous mammals as seed dispersers in Mediterranean dehesa. Oecologia 104: 246–255.

Mason, H. L. 1946a. The edaphic factor in narrow endemism. I. The nature of environmental influences. Madroño 8: 209–226.

Mason, H. L. 1946b. The edaphic factor in narrow endemism. II. The geographic occurrence of plants of highly restricted patterns of distribution. Madroño 8: 241–257.

May, R. M., J. H. Lawton, and N. E. Stork. 1995. Assessing extinction rates. Pages 1–24 in J. H. Lawton and R. M. May, eds., Extinction Rates. Oxford University Press, New York.

Mayer, M. S., and P. S. Soltis. 1994. A chloroplast DNA phylogeny of the *Strepthanthus glandulosus* complex (Cruciferae). Syst. Bot. 19: 557–574.

Mayer, S. S., D. Charlesworth, and B. Meyers. 1996. Inbreeding depression in four populations of *Collinsia heterophylla* Nutt. (Scrophulariaceae). Evolution 50: 879–891.

Mayr, E. 1942. Systematics and the Origin of Species. Columbia University Press, New York.

Mayr, E. 1963. Animal Species and Evolution. Belknap Press, Cambridge, Mass.

Mayr, E. 1969. The biological meaning of species. Biol. J. Linn. Soc. 1: 311–320.

Mays, E. 1970. Populations, Species and Evolution. Harvard University Press, Cambridge.

Mayr, E. 1982a. Speciation and macroevolution. Evolution 36: 1119–1132.

Mayr, E. 1982b. The Growth of Biological Thought. Harvard University Press, Cambridge, Mass.

McCarthy, E. M., M. A. Asmussen, and W. W. Anderson. 1995. A theoretical assessment of recombinational speciation. Heredity 74: 502–509.

McCauley, D. E. 1993. Genetic consequences of extinction and recolonization in fragmented habitats. Pages 217–233 in P. M. Kareiva, J. G. Kingsolver, and R. B. Huey, eds., Biotic Interactions and Global Change. Sinauer Associates, Sunderland, Mass.

McCauley, D. E. 1994. Contrasting the distribution of chloroplast DNA and allozyme polymorphism among local populations of *Silene alba*: implications for studies of gene flow in plants. Proc. Natl. Acad. Sci. USA. 91: 8127–8131.

McCauley, D. E. 1995. The use of chloroplast DNA polymorphism in studies of gene flow in plants. Trends Ecol. Evol. 10: 198–202.

McCauley, D. E., J. Raveill, and J. Antonovics. 1995. Local founding events as determinants of genetic structure in a plant metapopulation. Heredity 75: 630–636.

McClenaghan, L. R., Jr., and A. C. Beauchamp. 1986. Low genic differentiation among isolated populations of the California fan palm (*Washingtonia filifera*). Evolution 40: 315–322.

McClintock, B. 1984. The significance of responses of the genome to challenge. Science 226: 792–801.

McDade, L. A. 1995. Species concepts and problems in practice: insight from botanical monographs. Syst. Bot. 20: 606–622.

McDonald, J. F. 1995. Transposable elements: possible catalysts of organismic evolution. Trends Ecol. Evol. 10: 123–126.

McKinney, M. L. 1997. Extinction vulnerability and selectivity: combining ecological and paleontological views. Ann. Rev. Ecol. Syst. 28: 495–516.

McLeod, M. J., S. I. Guttman, W. H. Eshbaugh, and R. E. Rayle. 1983. An electrophoretic study of evolution in *Capsicum* (Solanaceae). Evolution 37: 562–574.

McNeill, C. I., and S. K. Jain. 1983. Genetic differentiation studies and phylogenetic inference in the plant genus *Limnanthes* (section *Inflexae*). Theor. Appl. Genet. 66: 257–269.

McNeilly, T., and J. Antonovics. 1968. Evolution in closely adjacent populations. IV. Barriers to gene flow. Heredity 23: 205–218.

McPeek, M. A. 1996. Trade-offs, food web structure, and the coexistence of habitat specialists and generalists. Amer. Natur. 148 (suppl): 124s–138s.

Menges, E. S. 1990. Population viability analysis for an endangered species. Conserv. Biol. 4: 52–62.

Menges, E. S. 1991. Seed germination percentage increases with population size in a fragmented prairie species. Conserv. Biol. 5: 158–164.

Menges, E. S., and S. C. Gawler. 1986. Four-year changes in population size of the endemic Furbish's lousewort: implications for endangerment and management. Nat. Areas J. 6: 6–17.

Menges, E. S., and R. W. Dolan. 1998. Demographic viability of populations of *Silene regia* in midwestern prairies: relationships with fire management, genetic variation, geographic location, population size and isolation. J. Ecol. 86: 63–78.

Merrell, D. J. 1968. The evolutionary role of dominant genes. Genetic Lectures (Oregon) 1: 167–194.

Michener, C. D. 1970. Diverse approaches to systematics. Evol. Biol. 4: 1–38.

Millar, C. I. 1983. A steep cline in *Pinus muricata*. Evolution 37: 311–319.

Milligan, B. 1986. Punctuated evolution induced by ecological change. Amer. Natur. 127: 522–532.

Mills, L. S., and P. E. Smouse. 1994. Demographic consequences of inbreeding in remnant populations. Amer. Natur. 144: 412–431.

Mindell, D. P., J. W. Sites, Jr., and D. Grauer. 1990. Mode of allozyme evolution: increased genetic distance associated with speciation events. J. Evol. Biol. 3: 125–131.

Mitchell-Olds, T. 1986. Quantitative genetics of survival and growth in *Impatiens capensis*. Evolution 40: 107–116.

Mitchell-Olds, T. 1996a. Genetic constraints on life-history evolution: quantitative trait loci influencing growth and flowering in *Arabidopsis thaliana*. Evolution 50: 140–145.

Mitchell-Olds, T. 1996b. Pleiotropy causes long-term genetic constraints on life history evolution in *Brassica napa*. Evolution 50: 1849–1858.

Mitchell-Olds, T., and J. L. Rutledge. 1986. Quantitative genetics in natural plant populations: a review of the theory. Amer. Natur. 127: 379–402.

Moll, R. H., J. H. Lonnquist, J. Velez Fortuno, and E. C. Johnson. 1965. The relationship of heterosis and genetic divergence in maize. Genetics 52: 139–144.

Monasterio, M., and L. Sarmiento. 1991. Adaptive radiation of *Espeletia* in the cold Andean tropics. Trends Ecol. Evol. 6: 387–391.

Moody, M. E., and R. N. Mack. 1988. Controlling the spread of plant invasions: the importance of nascent foci. J. Appl. Ecol. 25: 1009–1021.

Moore, D. M., and H. Lewis. 1965. The evolution of self-pollination in *Clarkia xantiana*. Evolution 19: 104–114.

Moore, W. S., and J. T. Price. 1993. Nature of selection in the northern flicker hybrid zone and its implications for speciation theory. Pages 196–225 in R. G. Harrison, ed., Hybrid Zones and the Evolutionary Process. Oxford University Press, New York.

Moran, G. F., and S. D. Hopper. 1983. Genetic diversity and insular population structure of the rare granite rock species *Eucalyptus caesia* Benth. Austral. J. Bot. 31: 161–172.

Moran, G. F., and S. D. Hopper. 1987. Conservation of the genetic resources of rare and widespread eucalypts in remnant vegetation. Pages 151–162 in D. A. Saunders, G. W. Arnold, A. A. Buridge, and A. J. M. Hopkins, eds., The Role of Remnants of Native Vegetation. Surrey Beatty, Sydney.

Morley, B. 1972. The distribution and variation of some gesneriads on Caribbean Islands. Pages 239–257 in D. H. Valentine, ed., Taxonomy, Phytogeography and Evolution. Academic Press, New York.

Morris, W. F. 1993. Predicting the consequences of plant spacing and biased movement for pollen dispersal by honeybees. Ecology 74: 493–500.

Mosquin, T. 1964. Chromosomal repatterning in *Clarkia rhomboidea* as evidence for post-Pleistocene changes in distribution. Evolution 18: 12–25.

Murawski, D. A., and J. L. Hamrick. 1991. The effect of density of flowering individuals on the mating systems of nine tropical tree species. Heredity 67: 167–174.

Murawski, D. A., J. L. Hamrick, S. P. Hubbell, and R. B. Foster. 1990. Mating systems of two bombacaceous trees of a Neotropical forest. Oecologia 82: 501–506.

Murphy, J. P. 1981. *Senecio × albescens* Burbidge & Colgan at Killiney Co., Dublin: a seventy-eight year old population. Watsonia 13: 303–311.

Myers, N. 1988. Threatened biotas: "hotspots" in tropical forests. Environmentalist 8: 1–20.

Nagy, E. S.. 1997a. Selection for native characters in hybrids between two locally adapted plant subspecies. Evolution 51: 1469–1480.

Nagy, E. S. 1997b. Frequency-dependent seed production and hybridization rates: implications for gene flow between locally adapted plant populations. Evolution 51: 703–714.

Nagy, E. S., and K. J. Rice. 1997. Local adaptation in two subspecies of an annual plant: implications for migration and gene flow. Evolution 51: 1079–1089.

Nason, J. D., E. A. Herre, and J. L. Hamrick. 1997. The breeding structure of a tropical keystone plant resource. Nature 391: 685–687.

Nei, M. 1987. Molecular Evolutionary Genetics. Columbia University Press, New York.

Nei, M., T, Maruyama, and R. Chakraborty. 1975. The bottleneck effect and genetic variability in populations. Evolution 29: 1–10.

Neigel, J. E., and J. C. Avise. 1986. Phylogenetic relationships of mitochondrial DNA under various demographic models of speciation. Pages 515–535 in S. Karlin and E. Nevo, eds., Evolutionary Processes and Theory. Academic Press, New York.

Nesom, G. L. 1983. *Galax* (Diapensiaceae): geographic variation in chromosome number. Syst. Bot. 8: 1–14.

Nevo, E. 1989. Modes of speciation: the nature and role of peripheral isolates in the origin of species. Pages 205–236 in L. V. Giddings, K. Y. Kanishiro, and W. W. Anderson, eds., Genetics, Speciation and the Founder Principle. Oxford University Press, New York.

New, J. K. 1978. Change and stability of clines in *Spergula arvensis* L. (corn spurry) after 20 years. Watsonia 12: 137–143.

Newman, D., and D. Pilson. 1997. Increased probability of extinction due to decreased genetic effective size: experimental populations of *Clarkia pulchella*. Evolution 51: 354–362.

Nicholls, M. K., and T. McNeilly. 1982. The possible polyphyletic origin of copper tolerance in *Agrostis tenuis* (Gramineae). Plant Syst. Evol. 140: 109–117.

Nichols, R. A., and G. M. Hewitt. 1994. The genetic consequences of long distance dispersal during colonization. Heredity 72: 312–317.

Niklas, K. J., B. H. Tiffney, and A. H. Knoll. 1983. Patterns in vascular plant diversification. Nature 303: 614–616.

Nilsson, Ö., and U. Wästljung. 1987. Seed predation and cross-pollination in mast-seeding beech (*Fagus sylvatica*) patches. Ecology 68: 260–265.

Nixon, K. C., and Q. D. Wheeler. 1990. An amplification of the phylogenetic species concept. Cladistics 6: 211–233.

Nobs, M. A., and W. M. Hiesey. 1957. Studies on differential selection in *Mimulus*. Carnegie Inst. Washington Yearbook 56: 291–292.

Nobs, M. A., and W. M. Hiesey. 1958. Performance of *Mimulus* races and their hybrids in contrasting environments. Carnegie Inst. Washington Yearbook 57: 270–272.

Nordenskiold, H. 1971. Hybridization experiments in the genus *Luzula*. IV. Studies with taxa of the *campestris-multiflora* complex from the Northern and Southern Hemispheres. Hereditas 68: 47–60.

Norrington-Davies, J. 1972. Diallel analysis of competition between some barley species and their hybrids. Euphytica 21: 292–308.

Novak, S. J., D. E. Soltis, and P. S. Soltis. 1991. Ownbey's Tragopogons: 40 years later. Amer. J. Bot. 78: 1586–1600.

Novacek, M. 1984. Evolutionary stasis in the elephant-shrew, *Rhynchocyon*. Pages 4–22 in N. Eldredge and S. M. Stanley, eds., Living Fossils. Springer-Verlag, New York.

O'Brien, T. A., W. J. Whittington, and P. Slack. 1967. Competition between perennial ryegrass, meadow fescue, and their natural hybrid: variation in growth rates and the proportion of each species with time. J. Appl. Ecol. 4: 501–512.

Odasz, A. M., and O. Savolainen. 1996. Genetic variation in populations of the arctic perennial *Pedicularis dasyantha* (Scrophulariaceae), on Svalbard, Norway. Amer. J. Bot. 83: 1379–1385.

Oka, H. I. 1957. Complementary lethal genes in rice. Japan J. Genet. 32: 83–87.

Olesen, J. M., and S. K. Jain. 1994. Fragmented plant populations and their lost interactions. Pages 417–426 in V. Loeschcke, J. Tomiuk, and S. K. Jain, eds., Conservation Genetics. Birkhäuser Verlag, Basel, Switzerland.

Ollerton, J. 1996. Reconciling ecological processes with phylogenetic patterns: the apparent paradox of plant-pollinator systems. J. Ecol. 767–769.

Olivieri, I., D. Couvet, and P-H. Gouyon. 1990. The genetics of transient populations: research at the metapopulation level. Trends Ecol. Evol. 5: 207–210.

Olmstead, R. G. 1995. Species concepts and plesiomorphic species. Syst. Bot. 20: 623–630.

Oostermeijer, J. G. B., R. G. M. Altenburg, and H. C. M. Den Nijs. 1995. Effects of outcrossing distance and selfing on fitness components in the rare *Gentiana pneumonanthe* (Gentianaceae). Acta Bot. Neerl. 44: 257–268.

Oostermeijer, J. G. B., M. W. van Eijck, and H. C. M. den Nijs. 1994. Offspring fitness in relation to population size and genetic variation in the rare perennial plant species *Gentiana pneumonanthe* (Gentianaceae). Oecologia 97: 289–296.

Orians, G. H. 1997. Evolved consequences of rarity. Pages 190–208 in W. E. Kunin and K. J. Gaston, eds., The Biology of Rarity. Chapman & Hall, London.

Ornduff, R. 1966. A biosystematic survey of the goldfield genus *Lasthenia*. University California Publ. Bot. 40: 1–92.

Orr, H. A., and J. A. Coyne. 1992. The genetics of adaptation: a reassessment. Amer. Natur. 140: 725–742.

Orr, H. A., and L. H. Orr. 1996. Waiting for speciation: the effect of population subdivision on the time to speciation. Evolution 50: 1742–1749.

Otte, D., and J. A. Endler. eds. 1989. Speciation and Its Consequences. Sinauer Associates, Sunderland, Mass.

Ouborg, N. J. 1993. Isolation, population size, and extinction: the classical and metapopulation approaches applied to vascular plants along the Dutch Rhine-system. Oikos 66: 298–308.

Ouborg, N. J., and R. van Treuren. 1994. The significance of genetic erosion in the process of extinction. IV. Inbreeding load and heterosis in relation to population size in the mint *Salvia pratensis*. Evolution 48: 996–1008.

Ouborg, N. J., R. van Treuren, and J. M. M. van Damme. 1991. The significance of genetic erosion in the process of extinction. II. Morphological variation and fitness components in populations of varying size of *Salvia pratensis* L. and *Scabiosa columbaria* L. Oecologia 86: 359–367.

Ownbey, M. 1950. Natural hybridization and amphiploidy in the genus *Tragopogon*. Amer. J. Bot. 40: 37: 487–499.

Paige, K. N., W. C. Capman, and P. Jennetten. 1991. Mitochondrial inheritance patterns across a cottonwood hybrid zone: cytonuclear disequilibria and hybrid zone dynamics. Evolution 45: 1360–1369.

Pamilo, P., and M. Nei. 1988. Relationships between gene trees and species trees. Molec. Biol. Evol. 3: 254–259.

Panetsos, C., and H. G. Baker. 1968. The origin of variation in "wild" *Raphanus sativus* (Cruciferae) in California. Genetica 38: 243–274.

Parks, C. R., N. G. Miller, J. F. Wendel, and K. M. McDougal. 1983. Genetic divergence within the genus *Liriodendron* (Magnoliaceae). Ann. Missouri Bot. Gard. 70: 658–666.

Parks, C. R., and J. F. Wendel. 1990. Molecular divergence between Asian and North American species of *Liriodendron* (Magnoliaceae) with implications for interpretations of fossil floras. Amer. J. Bot. 77: 1243–1256.

Parks, C. R., J. F. Wendel, M. M. Sewell, and Y-L. Qiu. 1994. The significance of allozyme variation and introgression in the *Liriodendron tulipifera* complex (Magnoliaceae). Amer. J. Bot. 81: 878–889.

Paterniani, E. 1969. Selection for reproductive isolation between two populations of maize, *Zea mays* L. Evolution 23: 534–547.

Paterson, A. H. 1995. Molecular dissection of quantitative traits: progress and prospects. Genome Res. 5: 321–333.

Paterson, A. H., Y-R. Lin, Z. Li, K. F. Schertz, J. F. Doebley, S. R. M. Pinson, S-C. Liu, J. W. Stansel, and J. E. Irvine. 1995. Convergent domestication of cereal crops by independent mutations at corresponding genetic loci. Science 269: 1714–1718.

Paterson, A. H., K. F. Schertz, Y-R. Lin, S-C. Liu and Y-L. Chang. 1995. The weediness of wild plants: molecular analysis of genes influencing dispersal and persistence of johnsongrass, *Sorghum halepense* (L.) Pers. Proc. Natl. Acad. Sci. USA. 92: 6127–6131.

Paterson, H. E. H. 1978. More evidence against speciation by reinforcement. South African J. Sci. 78: 369–371.

Paterson, H. E. H. 1985. The recognition concept of species. Pages 21–29 in E. E. Vrba, ed., Species and Speciation. Transvaal Museum Monogr. No. 4, Pretoria.

Paterson, H. E. H. 1986. Environment and species. South African J. Sci. 82: 62–65.

Paterson, H. E. H. 1993. Animal species and sexual selection. Pages 209–228 in Evolutionary Patterns and Processes. Linnean Soc. London.

Patterson, W. A. III, and A. E. Backman. 1988. Fire and disease history of forests. Pages 603–632 in B. Huntley and T. Webb III, eds., Vegetation History. Kluwer Academic Publishers, Dordrecht, The Netherlands.

Pearson, P. N. 1998. Species and extinction asymmetries in paleontological phylogenies: evidence for evolutionary progress. Paleobiology 24: 305–335.

Pedersen, M. W., R. L. Hurst, M. D. Levin, and G. L. Stoker. 1969. Computer analysis of the genetic contamination of alfalfa seed. Crop Sci. 9: 1–4.

Pedersen, P. N., H. B. Johansen, and J. Jorgensen. 1961. Pollen spreading in diploid and tetraploid rye. II. Distance of pollen spreading and risk of intercrossing. Royal Vet. Agr. Coll. Ann. Yearbook, pp. 68–86.

Perring, S. H., and S. M. Walters. eds. 1976. Atlas of the British Flora, 2nd. ed. E. P. Publishing, East Ardsley.

Perrins, J., A. Fitter, and M. Williamson. 1993. Population biology and rates of invasion of three introduced *Impatiens* species in the British Isles. J. Biogeogr. 20: 33–44.

Petit, R. J., E. Pineau, B. Demesure, R. Bacilieri, A. Ducousso, and A. Kremer. 1997. Chloroplast DNA footprints of post-glacial recolonization by oaks. Proc. Natl. Acad. Sci. USA. 94: 9996–10001.

Pickering, R. A. 1983. The location of a gene for incompatibility between *Hordeum vulgare* L. and *H. bulbosum*. Heredity 51: 455–459.

Pigott, C. D. 1974. The response of plants to climate and climatic change. Pages 32–44 in F. Perring, ed., The Flora of a Changing Britain, Classey, Hampton.

Pigott, C. D. 1981. Nature of seed sterility and natural regeneration of *Tilia cordata* near its northern limit in Finland. Ann. Bot. Fenn. 18: 255–263.

Pleasants, J. M., and J. F. Wendel. 1989. Genetic diversity in a clonal narrow endemic, *Erythronium propullans*, and its widespread progenitor, *Erythronium album*. Amer. J. Bot. 76: 1136–1151.

Portnoy, S., and M. F. Willson. 1993. Seed dispersal curves: behavior of the tail of the distribution. Evol. Ecol. 7: 25–44.

Poschlod, P., and S. Bonn. 1998. Changing dispersal processes in the central European landscape since the last ice age: an explanation for the actual decrease of plant species richness in different habitats? Acta Bot. Neerl. 47: 27–44.

Powell, A. H., and G. V. N. Powell. 1987. Population dynamics of male euglossine bees in Amazonian forest fragments. Biotropica 19: 176–179.

Prat, D., C. Leger, and S. Bojovic. 1992. Genetic diversity among *Alnus glutinosa* (L.) Gaertn. populations. Acta Oecol. 13: 469–477.

Prazmo, W. 1965. Cytogenetic studies on the genus *Aquilegia*. III. Inheritance of traits that distinguish different complexes in the genus *Aquilegia*. Acta Soc. Bot. Polon. 34: 403–437.

Price, P. W., M. Westoby, and B. Rice. 1988. Parasite-mediated competition: some predictions and tests. Amer. Natur. 131: 544–555.

Prober, S. M., and A. H. D. Brown. 1994. Conservation of the grassy white box woodlands: population genetics and fragmentation of *Eucalyptus albens*. Conserv. Biol. 8(4): 1003–1013.

Pulliam, R. 1988. Sources, sinks, and population regulation. Amer. Natur. 132: 652–661.

Purdy, B. G., R. J. Bayer, and S. E. Macdonald. 1994. Genetic variation, breeding system evolution, and the conservation of the narrow sand dune endemic *Stellaria arenicola* and the widespread *S. longipes* (Caryophyllaceae). Amer. J. Bot. 81: 904–911.

Pysek, P. 1998. Is there a taxonomic pattern to plant invasions? Oikos 82: 282–294.

Pysek, P., and K. Prach. 1993. Plant invasions and the role of riparian habitats: a comparison of four species alien to central Europe. J. Biogeogr. 204: 413–420.

Pysek, P., K. Prach, and P. Smilauer. 1995. Relating invasion success to plant traits: an analysis of the Czech alien flora, Pages 39–60 in P. Pysek, K. Prach, M. Rejmánek and M. Wade, eds., Plant Invasions–General Aspects and Special Problems. SPB Academic Publishers, Amsterdam.

Quillet, M. C., N. Majidian, T. Griveau, H. Serieys, M. Tersac. M. Lorieus, and A. Bervillé. 1995. Mapping genetic factors controlling pollen viability in an interspecific cross in *Helianthus* section *Helianthus*. Theor. Appl. Genet. 91: 1195–1202.

Rabinowitz, D. 1981. Seven forms of rarity. Pages 205–217 in H. Synge, ed., The Biological Aspects of Rare Plant Conservation, Wiley, Chichester.

Rabinowitz, D., S. Cairns, and T. Dillon. 1986. Seven forms of rarity and their frequency in the flora of the British Isles. Pages 182–204 in M. E. Soulé, ed., Conservation Biology—The Science of Scarcity and Diversity. Sinauer Assoc., Sunderland, Mass.

Raijmann, L. E. L., N. C. Van Leeuwen, R. Kersten, J. G. B. Oostermeijer, H. C. M. Den Nijs, and S. B. J. Menken. 1994. Genetic variation and outcrossing rate in relation to population size in *Gentiana pneumonanthe* L. Conserv. Biol. 8(4): 1014–1026.

Rathcke, B. 1983. Competition and facilitation among plants for pollination. Pages 305–329 in L. Real, ed., Pollination Biology. Academic Press, New York.

Rathcke, B. J., and E. S. Jules. 1993. Habitat fragmentation and plant-pollinator interactions. Curr. Sci. 65: 273–276.

Rathcke, B. J., and L. Real. 1993. Autogamy and inbreeding depression in mountain laurel, *Kalmia latifolia* (Ericaceae). Amer. J. Bot. 80: 143–146.

Rattenbury, J. A. 1962. Cyclic hybridization as a survival mechanism in the New Zealand forest flora. Evolution 16: 348–363.

Rausher, M. D., and J. D. Fry. 1993. Effect of a locus affecting floral pigmentation in *Ipomoea purpurea* on female fitness components. Genetics 134: 1237–1247.

Raven, P. H. 1963. Amphitropical relationships in the floras of North and South America. Quart. Rev. Biol. 38: 151–177.

Raven, P. H. 1964. Catastrophic selection and edaphic endemism. Evolution 18: 336–338.

Raven, P. H. 1972. Evolution and endemism in the New Zealand species of *Epilobium*. Pages 259–274 in D. H. Valentine, ed., Taxonomy, Phytogeography and Evolution. Academic Press, New York.

Raven, P. H. 1977. Systematics and plant population biology. Syst. Bot. 1: 284–318.

Raven, P. H. 1978. Future directions of plant biology. Pages 461–481 in O. T. Solbrig et al., eds., Topics in Plant Population Biology. Columbia University Press, New York.

Raven, P. H. 1980. Hybridization and the nature of plant species. Canad. Bot. Assoc. Bull. Suppl. to Vol. 13: 3–10.

Raven, P. H. 1986. Modern aspects of the biological species in plants. Pages 11–29 in K. Iwatsuki et al., eds., Modern Aspects of Species. Tokyo University Press, Tokyo.

Raven, P. H. 1987. The scope of plant conservation problem world-wide. Pages 19–29 in O. Hamann, V. Heywood, and H. Synge, eds., Botanic Gardens and the World Conservation Strategy, Academic Press, London.

Raven, P. H. 1988. Our diminishing tropical forests. Pages 199–122 in E. O. Wilson and F. M. Peter, eds., Biodiversity. National Academy Press, Washington, D.C.

Raven, P. H., and D. I. Axelrod. 1978. Origin and relationships of the California flora. University California Publ. Bot. 72: 1–134.

Raven, P. H., B. Berlin, and D. E. Breedlove. 1971. The origins of taxonomy. Science 174: 1210–1213.

Raven, P. H., and D. Gregory. 1972. A revision of the genus *Gaura*. Mem. Torrey Bot. Club 23: 1–96.

Raven, P. H., D. W. Kyhos, D. E. Breedlove, and W. W. Payne. 1968. Polyploidy in *Ambrosia dumosa* (Compositae: Ambrosieae). Brittonia 20: 205–211.

Raven, P. H., and T. E. Raven. 1976. The Genus *Epilobium* (Onagraceae) in Australasia: A Systematic Evolutionary Study. Bascands, Ltd., Christchurch, New Zealand.

Reinartz, J. A. 1984. Life history variation of common mullein (*Verbascum thapsus*). 1. Latitudinal differences in population dynamics and timing of reproduction. J. Ecol. 72: 897–912.

Reinartz, J. A., and D. H. Les. 1994. Bottleneck-induced dissolution in self-incompatibility and breeding system consequences in *Aster furcatus* (Asteraceae). Amer. J. Bot. 81: 446–455.

Rejmánek, M. 1989. Invasibility of plant communities. Pages 369–388 in J. A. Drake, H. A. Mooney, F. di Castri, R. H. Grooves, F. J. Kruger, M. Rejmánek, and M. Williamson, eds., Biological Invasions–A Global Perspective. John Wiley & Sons, New York.

Rejmánek, M. 1996. Species richness and resistance to invasions. Pages 153–172 n G. H. Orians, R. Dirzo, and J. H. Cushman eds., Biodiversity and Ecosystem Processes in Tropical Forests. Springer-Verlag, New York.

Rejmánek, M., and D. M. Richardson. 1996. What attributes make some plant species more invasive? Ecology 77: 1655–1661.

Remington, C. L. 1968. Suture-zones of hybrid interaction between recently joined biotas. Evol. Biol. 2: 321–428.

Rhymer, J. M., and D. Simberloff. 1996. Extinction by hybridization. Ann. Rev. Ecol. Syst. 27: 83–109.

Richards, R. A., and N. Thurling. 1973. The genetics of self-incompatibility in *Brassica campestris* L. ssp *oleifera* Metzg. II. Genotypic and environmental modification of S locus control. Genetica 44: 439–453.

Rick, C. M. 1963. Barriers to interbreeding in *Lycopersicon peruvianum*. Evolution 17: 216–232.

Rick, C. M. 1988. Evolution of mating systems in cultivated plants. Pages 133–147 in L. D. Gottlieb and S. K. Jain, eds., Plant Evolutionary Biology. Chapman and Hall, New York.

Ricklefs, R. E. 1989. Speciation and diversity: the integration of local and regional processes. Pages 599–622 in D. Otte and J. A. Endler, eds., Speciation and Its Consequences. Sinauer Associates, Sunderland, Mass.

Ricklefs, R. E., and R. E. Latham. 1992. Intercontinental correlation of geographical ranges suggests stasis in ecological traits of relict genera of temperate perennial herbs. Amer. Natur. 139: 1305–1321.

Rieseberg, L. H. 1991a. Homoploid reticulate evolution in *Helianthus* (Asteraceae): evidence from ribosomal genes. Amer. J. Bot. 78: 1218–1237.

Rieseberg, L. H. 1991b. Hybridization in rare plants: insights from case studies in *Cerocarpus* and *Helianthus*. Pages 171–181 in D. A. Falk and K. E. Holsinger, eds., Genetics and Conservation of Rare Plants. Oxford University Press, New York.

Rieseberg, L. H., and L. Brouillet. 1994. Are many plant species paraphyletic? Taxon 43: 21–32.

Rieseberg, L. H., R. Carter, and S. Zona. 1990. Molecular tests of the hypothesized hybrid origin of two diploid *Helianthus* species (Asteraceae). Evolution 44: 1498–1511.

Rieseberg, L. H., and D. Gerber. 1995. Hybridization in the Catalina Island mountain mahogany (*Cerocarpus traskiae*): RAPD evidence. Conserv. Biol. 9: 199–203.

Rieseberg, L. H., P. M. Peterson, D. E. Soltis, and C. R. Annable. 1987. Genetic divergence and isozyme number variation among four varieties of *Allium douglasii* (Alliaceae). Amer. J. Bot. 74: 1614–1624.

Rieseberg, L. H., B. Sinervo, C. R. Linder, M. C. Ungerer, and D. M. Arias. 1996. Role of gene interactions in hybrid speciation: evidence from ancient and experimental hybrids. Science 272: 741–745.

Rieseberg, L. H., C. VanFossen, and A. M. Desrochers. 1995. Hybrid speciation accompanied by genomic reorganization in wild sunflowers. Nature 375: 313–316.

Rieseberg, L. H., and J. F. Wendel. 1993. Introgression and its consequences in plants. Pages 70–109 in R. G. Harrison, ed., Hybrid Zones and the Evolutionary Process. Oxford University Press, New York.

Rieseberg, L. H., S. Zona, L. Aberbom, and T. D. Martin. 1989. Hybridization in the island endemic, Catalina mahogany. Conserv. Biol. 3: 52–58.

Rodriguez, D. J. 1996. A model for the establishment of polyploidy in plants. Amer. Natur. 147: 33–46.

Rodríquez, P. A. M., and P. L. Pére de Paz. 1997. Flora terrestre de Canarias y su biodiversidad. Pages 177–189 in P. L. Pérez de Paz, ed., Ecosistemas Insulares Canarios. Santa Cruz de Tenerife.

Roelofs, D., J. Van Velzen, P. Kuperus, and K Bachmann. 1997. Molecular evidence for an extinct parent of the tetraploid species *Microseris acuminata* and *M. campestris* (Asteraceae, Lactuceae). Molec. Ecol. 6: 641–649.

Rothera, S. L., and A. J. Davy. 1986. Polyploidy and habitat differentiation in *Deschampsia cespitosa*. New Phytol. 102: 449–467.

Rouhani, S., and N. H. Barton. 1987. Speciation and the "shifting balance" in continuous populations. Theor. Pop. Biol. 31: 465–492.

Roy, J. 1990. In search of the characteristics of plant invaders. Pages 335–352 in A. J. di Castri, A. J. Hansen, and M. Debushe, eds., Biological Invasions in Europe and the Mediterranean Basin. Kluwer, Dordrecht, Netherlands.

Russell, J. R., and D. A. Levin. 1988. Competitive relationships of *Oenothera* species with different recombination systems. Amer. J. Bot. 75: 1175–1180.

Sakai, A. K., W. L. Wagner, D. M. Ferguson, and D. R. Herbst. 1995. Biogeographical and ecological correlates of dioecy in the Hawaiian flora. Ecology 76: 2530–2543.

Salas-Pascual, M., J. R. Acebes-Ginoves, and M. Del Acro-Aguilar. 1993. *Arbutus × androsterilis*, a new interspecific hybrid between *A. canariensis* and *A. unedo* from the Canary Islands. Taxon 42: 789–792.

Salisbury, E. J. 1953. A changing flora as shown in the study of weeds of arable lands and waste places. Pages 130–139 in J. E. Lousley, ed., The Changing Flora of Britain. Botanical Society of the British Isles, London.

Sano, Y., and F. Kita. 1978. Reproductive barriers distributed in *Melilotus* species and their genetic bases. Canad. J. Genet. Cytol. 20: 275–289.

Santos-Guerra, A., J. Francisco-Ortega, and E. Feria. 1993. Contributions to the knowledge of *Argyranthemum* Webb ex Sch. Bip (Compositae) in the Canary Islands. P. 27 in Abstracts First Symposium Fauna and Flora of Atlantic Islands. Museu Municipaldo Funchel, Funcel, Madeira.

Sauer, J. 1990. Allopatric speciation: deduced but not detected. J. Biogeogr. 17: 1–3.

Schat, H., E. Kuiper, W. M. Ten Bookum, and R. Vooijs. 1993. A general model for the genetic control of copper tolerance in *Silene vulgaris*: evidence from crosses between plants from different tolerant populations. Heredity 70: 142–147.

Schat, H., R. Vooijs, and E. Kuiper. 1996. Identical major gene loci for heavy metal tolerances that have evolved independently in different local populations and subspecies of *Silene vulgaris*. Evolution 50: 1888–1895.

Scheller, G., M. T. Conkle, and L. Griswald. 1985. Local differentiation among Mediterranean populations of Aleppo pine in their isozymes. Silvae Genet. 35: 11–19.

Schemske, D. W., and C. C. Horvitz. 1989. Temporal variation in selection on a floral character. Evolution 43: 461–465.

Schemske, D. W., B. C. Husband, M. H. Ruckelshaus, C. Goodwillie, I. M. Parker, and J. G. Bishop. 1994. Evaluating approaches to the conservation of rare and endangered plants. Ecology 75: 584–606.

Schlichting, C. D., and D. A. Levin. 1986. Phenotypic plasticity: an evolving plant character. Biol. J. Linn. Soc. 29: 37–47.

Schluter, D. 1998. Ecological causes of speciation. Pages 114–129 in D. J. Howard and S. H. Berlocher, eds., Endless Forms-Species and Speciation. Oxford University Press, New York.

Schnabel, A. F. 1988. Population genetic structure and gene flow in *Gleditsia triacanthos* L. Ph.D. Dissertation. University of Kansas, Lawrence.

Schmidt, K., and D. A. Levin. 1985. The comparative demography of reciprocally sown populations of *Phlox drummondii* Hook. Survivorships, fecundities, and finite rates of increase. Evolution 39: 396–404.

Schmitz, U. K. 1988. Dwarfism and male sterility in interspecific hybrids of *Epilobium*. 1. Expression of plastid genes and structure of the plastome. Theor. Appl. Genet. 75: 350–356.

Schmitz, U. K., and G. Michaelis. 1988. Dwarfism and male sterility in interspecific hybrids of *Epilobium.* 2. Expression of mitochondrial genes and the structure of mitochondrial DNA. Theor. Appl. Genet. 76: 565–569.

Schoen, D. J. 1977. Morphological, phenological, and pollen distribution evidence of autogamy and xenogamy in *Gilia achilleaefolia* (Polemoniaceae). Syst. Bot. 2: 280–286.

Schoen, D. J. 1982. The breeding system of *Gilia achilleaefolia*: variation in floral characteristics and outcrossing rate. Evolution 36: 352–360.

Schoen, D. J., and S. C. Stewart. 1987. Variation in male fertilities and pairwise mating probabilities in *Picea glauca*. Genetics 116: 141–152.

Schoen, D. J., M. O. Johnston, A-M. L'Heureux, and J. V. Marsolais. 1997. Evolutionary history of the mating system in *Amsinckia* (Boraginaceae). Evolution 51: 1090–1099.

Schonewald-Cox, C., and M. Buechner. 1991. Housing viable populations in protected habitats: the value of a course-grained geographic analysis of density patterns and available habitat. Pages 213–226 in A. Seitz and V. Loeschcke, eds., Species Conservation: A Population Biological Approach. Birkhäuser Verlag, Berlin.

Schwaegerle, K. E., and B. A. Schaal. 1979. Genetic variability and founder effect in the pitcher plant *Sarracenia purpurea* L. Evolution 33: 1210–1218.

Schwaegerle, K. E., K. Garbutt, and F. A. Bazzaz. 1986. Differentiation among nine populations of *Phlox*. I. Electrophoretic and quantitative variation. Evolution 40: 506–517.

Schwaegerle, K. E., and D. A. Levin. 1991. Quantitative genetics of fitness traits in a wild population of *Phlox*. Evolution 45: 169–177.

Shaffer, M. 1981. Minimum population sizes for conservation. Bioscience 31: 131–134.

Shaffer, M. 1987. Minimum viable populations: coping with uncertainty. Pages 69–86 in M. Soulé, ed., Viable Populations for Conservation. Cambridge University Press, Cambridge.

Sharsmith, H. K. 1961. The genus *Hesperolinon* (Linaceae). University California Publ. Bot. 32: 235–314.

Shaw, K. L. 1998. Species and the diversity of natural groups. Pages 44–56 in D. J. Howard and S. H. Berlocher, eds., Endless Forms–Species and Speciation. Oxford University Press, New York.

Shaw, M. W. 1995. Simulation of population expansion and spatial pattern when individuals dispersal distribution do not decline exponentially with distance. Proc. Royal Soc. Lond. B. 259: 243–248.

Sheely, D. L., and T. R. Meagher. 1996. Genetic diversity in Micronesian Island populations of the tropical tree *Campnosperma brevipetiolata* (Anacardiaceae). Amer. J. Bot. 83: 1571–1579.

Sherman, M. 1946. Karyotype evolution: a cytogenetic study of seven species and six interspecific hybrids in *Crepis*. University California Publ. Bot. 18: 369–408.

Sherman-Broyles, S. L., J. P. Gibson, J. L. Hamrick, M. A. Bucher, and M. J. Gibson. 1992. Comparisons of allozyme diversity among rare and widespread *Rhus* species. Syst. Bot. 17: 551–559.

Shigesada, N., K. Kawasaki, and Y. Takeda. 1995. Modeling stratified diffusion in biological invasions. Amer. Natur. 146: 229–251.

Sieber, V. K., and B. G. Murray. 1980. Spontaneous polyploids in marginal populations of *Alopecurus bulbosus* Gouan. (Poaceae). Bot. J. Linn. Soc. 81: 293–300.

Sih, A., and M. S. Baltus. 1987. Patch size, pollinator behavior, and pollinator limitation in catnip. Ecology 68: 1679–1690.

Silander, J. A. 1985. The genetic basis of ecological amplitude of *Spartina patens*. II. Variance and correlation analysis. Evolution 39: 1034–1052.

Simpson, G. G. 1961. Principles of Animal Taxonomy. Columbia University Press, New York.

Skellam, T. G. 1951. Random dispersal in theoretical populations. Biometrika 38: 196–218.

Slatkin, M. 1976. The spread of an advantageous allele in a subdivided population. Pages 767–780 in S. Karlin and E. Nevo, eds., Population Genetics and Ecology. Academic Press, New York.

Slatkin, M. 1977. Gene flow and genetic drift in a species subject to frequent local extinctions. Theor. Pop. Biol. 12: 253–262.

Slatkin, M. 1985. Gene flow in natural populations. Ann. Rev. Ecol. Syst. 16: 393–430.

Slatkin, M. 1987. Gene flow and the geographic structure of natural populations. Science 236: 787–792.

Small, E. 1972. Adaptation in *Clarkia*, Section Myxocarpa. Ecology 53: 808–818.

Smith, F. D. M., R. M. May, R. Pellew, T. H. Johnson, and K. R. Walter. 1993. How much do we know about the current extinction rate? Trends Ecol. Evol. 8: 375–378.

Smith, H. H. 1950. Developmental restrictions on recombination in *Nicotiana*. Evolution 4: 202–211.

Smith, H. H. 1968. Recent cytogenetic studies in the genus *Nicotiana*. Adv. Genet. 14: 1–54.

Smith, H. H., and K. Daly. 1959. Discrete populations derived by interspecific hybridization and selection in *Nicotiana*. Evolution 13: 476–487.

Smith, J. F., C. C. Burke, and W. L. Wagner. 1996. Interspecific hybridization in natural populations of *Cyrtandra* (Gesneriaceae) on the Hawaiian Islands–evidence from RAPD markers. Plant Syst. Evol. 200: 61–77.

Smith, J. M. B. 1981. Colonist ability, altitudinal range and origins of the flora of Mt. Field, Tasmania. J. Biogeogr. 8: 249–261.

Smyth, C. A., and J. L. Hamrick. 1984. Variation in estimates of outcrossing in musk thistle populations. J. Hered. 75: 303–307.

Snogerup, S. 1967. Studies in the Aegean flora. IX. *Erysimum* sect. Cheiranthus. B. Variation and evolution in the small population system. Opera Botanica 14: 1–86.

Sokal, R. R., and P. Menozzi. 1982. Spatial autocorrelation of HLA frequencies in Europe support demic diffusion of early farmers. Amer. Natur. 119: 1–17.

Soltis, D. E. 1981. Allozyme variability in *Sullivantia* (Saxifragaceae). Syst. Bot. 7: 26–34.

Soltis, D. E. 1984. Autopolyploidy in *Tolmiea menziesii* (Saxifragaceae). Amer. J. Bot. 71: 1171–1174.

Soltis, D. E. 1985. Allozymic differentiation among *Heuchera americana, H. parviflora, H. pubescens*, and *H. villosa* (Saxifragaceae). Syst. Bot. 10: 193–198.

Soltis, D. E., and J. J. Doyle. 1987. Ribosomal RNA gene variation in diploid and tetraploid *Tolmiea menziesii* (Saxifragaceae). Biochem. Syst. Ecol. 15: 75–78.

Soltis, P., G. Plunkett, S. Novak, and D. Soltis. 1995. Genetic variation in *Tragopogon* species: additional origins of the allotetraploids *T. mirus* and *T. miscellus*. Amer. J. Bot. 82: 1329–1341.

Soltis, D. E., and P. S. Soltis. 1989. Genetic consequences of autopolyploidy in *Tolmiea* (Saxifragaceae). Evolution 43: 586–594.

Soltis, D. E., and P. S. Soltis. 1993. Molecular data and the dynamic nature of polyploidy. Critical Reviews in Plant Sciences 12: 243–273.

Soltis, D. E., and P. S. Soltis. 1995. The dynamic nature of polyploid genomes. Proc. Natl. Acad. Sci. USA. 92: 8089–8091.

Soltis, D. E., P. S. Soltis, and B. D. Ness. 1989. Chloroplast DNA variation and multiple origins of autopolyploidy in *Heuchera micrantha* (Saxifragaceae). Evolution 43: 650–656.

Song, K., P. Liu, K. Tang, and T. C. Osborn. 1995. Rapid genome change in synthetic polyploids of *Brassica* and its implications for polyploid evolution. Proc. Natl. Acad. Sci. USA. 92: 7719–7723.

Soulé, M. E. 1980. Thresholds for survival: maintaining fitness and evolutionary potential.

Pages 151–169 in M. E. Soulé and B. A. Wilcox, eds., Conservation Biology: an Evolutionary Perspective. Sinauer Associates. Sunderland, Mass.

Soulé, M., and B. A. Wilcox. 1980. Conservation Biology: An Evolutionary-ecological Perspective. Sinauer, Sunderland, Mass.

Sowig, P. 1989. Effects of flowering plant's patch size on species composition of pollinator communities, foraging strategies, and resource partitioning in bumblebees (Hymenoptera: Apidae). Oecologia 78: 550–558.

Spencer, H. G., B. H. McArdle, and D. M. Lambert. 1986. A theoretical investigation of speciation by reinforcement. Amer. Natur. 128: 241–262.

Squillace, A. E. 1974. Average genetic correlations among offspring from open-pollinated forest trees. Silvae Genet. 23: 149–156.

Squillace, A. E., and J. R. Krause. 1963. The degree of natural selfing in slash pine as estimated from albino frequencies. Silvae Genet. 12: 46–50.

Stace, C. A. 1975. Hybridization and the Flora of the British Isles. Academic Press, New York.

Stanley, S. M. 1975. A theory of evolution above the species level. Proc. Natl. Acad. Sci. USA. 72: 646–650.

Stanley, S. M. 1978. Chronospecies' longevities, the origin of genera, and the punctuational model of evolution. Paleobiology 4: 26–40.

Stanley, S. M. 1979. Macroevolution—Pattern and Process. W. H. Freeman, San Francisco.

Stanley, S. M. 1986. Population size, extinction, and speciation: the fission effect in Neogene Bivalva. Paleobiology 12: 89–110.

Stanley, S. M. 1990. The general correlation between rate of speciation and rate of extinction: fortuitous causal linkages. Pages 103–127 in R. M. Ross and W. D. Allmon, eds., Causes of Evolution. University of Chicago Press, Chicago.

Stebbins, G. L. 1950. Variation and Evolution in Plants. Columbia University Press, New York.

Stebbins, G. L. 1958. The inviability, weakness, and sterility of interspecific hybrids. Adv. Genet. 9: 147–215.

Stebbins, G. L. 1959. Genes, chromosomes and evolution. Pages 258–290 in W. Turrill, ed., Vistas in Botany. Pergamon, London

Stebbins, G. L. 1969a. Comments on the search for a "perfect system." Taxon: 18: 357–359.

Stebbins, G. L. 1969b. The significance of hybridization for plant taxonomy and evolution, Taxon 18: 26–35.

Stebbins, G. L. 1971. Chromosomal Evolution in Higher Plants. Addison-Wesley Publ. Co., Reading, Mass.

Stebbins, G. L. 1974. Flowering Plants—Evolution Above the Species Level. Belknap Press, Cambridge, Mass.

Stebbins, G. L. 1980. Rarity of plant species: a synthetic viewpoint. Rhodora 82: 77–86.

Stebbins, G. L. 1982. Plant speciation. Pages 21–39 in C. Barigozzi, ed., Mechanisms of Speciation. Alan R. Liss, New York.

Stebbins, G. L. 1987. Species concepts: semantics and actual situations. Biology & Philosophy 2: 198–203.

Stebbins, G. L. 1989. Plant species and the founder principle. Pages 113–125 in L. V. Giddings, K. Y. Kaneshiro, and W. W. Anderson, eds., Genetics, Speciation and the Founder Principle. Oxford University Press, New York.

Stebbins, G. L., and K. Daly. 1961. Changes in the variation pattern of a hybrid population of Helianthus over an eight-year period. Evolution 15: 60–71.

Steinhoff, R. J., D. G. Joyce, and L. Fins. 1983. Isozyme variation in Pinus monticola. Canad. J. Forest Res. 13: 1122–1132.

Stephens, S. G. 1950. The genetics of "corky." II. Further studies of its genetic basis in relation to the general problem of interspecific isolating mechanisms. J. Genet. 50: 9–20.

Stoller, E. W. 1973. Effect of minimum soil temperature on differential distribution of *Cyperus rotundifolia* and *C. esculentus* in the United States. Weed Res. 13: 209–217.

Stoutamire, W. P. 1977. Chromosome races of *Gaillardia pulchella* (Asteraceae-Heliantheae). Brittonia 29: 297–309.

Stowe, K. A. 1998. Experimental evolution of resistance in *Brassica rapa*: correlated response of tolerance in lines selected for glucosinolate content. Evolution 52: 703–712.

Strid, A. 1970. Studies in the Aegean flora. XVI. Biosystematics of the *Nigella arvensis* complex with special reference to the problem of non-adaptive radiation. Opera Botanica 33: 1–118.

Stuessy, T. D., D. J. Crawford, C. Marticorina, and R. Rodriguez. 1998. Island biogeography of angiosperms of the Juan Fernandez Islands. Pages 121–138 in T. F. Stuessy and M. Ono, eds., Evolution and Speciation of Island Plants. Cambridge University Press, New York.

Stuessy, T. D., K. A. Foland, J. F. Sutter, and M. Silva O. 1984. Botanical and geological significance of potassium-argon dates from the Juan Fernandez Islands. Science 225: 49–51.

Sultan, S. E. 1987. Evolutionary implications of phenotypic plasticity in plants. Evol. Biol. 21: 127–178.

Sun, M., and F. R. Ganders. 1990. Outcrossing rates and allozyme variation in rayed and rayless morphs of *Bidens pilosa*. Heredity 64: 139–143.

Sytsma, K. J., and L. D. Gottlieb. 1986. Chloroplast DNA evolution and phylogenetic relationships in *Clarkia* Sect. *Peripetasma* (Onagraceae). Evolution 40: 1248–1261.

Sytsma, K. J., J. F. Smith, and L. D. Gottlieb. 1990. Phylogenetics in *Clarkia* (Onagraceae): restriction site mapping of chloroplast DNA. Syst. Bot. 15: 280–295.

Taberlet, P., L. Fumagalli, and A-G. Wust-Saucy. 1998. Comparative phylogeography and the postglacial colonization routes in Europe. Molec. Ecol. 7: 453–464.

Tadmor, Y., D. Zamir, and G. Ladizinsky. 1987. Genetic mapping of an ancient translocation in the genus *Lens*. Theor. Appl. Genet. 73: 883–892.

Taggart, J. B., S. F. McNally, and P. M. Sharp. 1990. Genetic variability and differentiation among founder populations of the pitcher plant (*Sarracenia purpurea* L.) in Ireland. Heredity 64: 177–183.

Tanksley, S. D., R. Bernatzky, N. L. Lapitan, and J. P. Prince. 1988. Conservation of gene repertoire but not gene order in pepper and tomato. Proc. Natl. Acad. Sci. USA. 85: 6419–6423.

Templeton, A. R. 1980. The theory of speciation via the founder principle. Genetics 94: 1011–1038.

Templeton, A. R. 1981. Mechanisms of speciation—a population genetic approach. Ann. Rev. Ecol. Syst. 12: 23–48.

Templeton, A. R. 1986. Coadaptation and outbreeding depression. Pages 105–116 in M. E. Soulé, ed., Conservation Biology, The Science of Scarcity and Diversity. Sinauer Associates, Sunderland, Mass.

Templeton, A. R. 1989. The meaning of species and speciation: a genetic perspective. Pages 3–27 in D. Otte and J. A. Endler, eds., Speciation and Its Consequences. Sinauer Associates, Sunderland, Mass.

Thebaud, C., and M. Debussche. 1991. Rapid invasion of *Fraxinus ornus* L. along the Herault River system in southern France: the importance of seed dispersal by water. J. Biogeogr. 18: 7–12.

Thomas, C. D. 1994. Extinction, colonization, and metapopulations; environmental tracking by rare species. Conserv. Biol. 8: 373–378.

Thompson, K., and A. Jones. 1999. Human population density and prediction of local plant extinction in Britain. Conserv. Biol. 13: 185–189.

Togby, H. A. 1943. A cytological study of *Crepis fuliginosa, C. neglecta*, and their hybrid, and its bearing on the mechanisms of phylogenetic reduction in chromosome number. J. Genet. 45: 67–111.

Townsend, C. E., and E. E. Remmenga. 1968. Inbreeding in the tetraploid alsike clover, *Trifolium hybridum*. Crop Sci. 8: 213–217.

Turner, I. M., K. S. Chua, J. S. Y. Ong, B. C. Soong, and H. T. W. Tan. 1996. A century of plant species loss from an isolated fragment of lowland tropical forest. Conserv. Biol. 10: 1229–1244.

Ungerer, M. C., S. J. Baird, J. Pan, and L. H. Rieseberg. 1998. Rapid hybrid speciation in sunflowers. Proc. Natl. Acad. Sci. USA. 95: 11757–11762.

van den Bosh, F., J. C. Zadocks, and J. A. J. Metz. 1988. Focus expansion in plant disease. II. Realistic parameter sparse models. Phytopathology 78: 59–64.

van der Pijl, L. 1961. Ecological aspects of flower evolution. II. Zoophilous flower classes. Evolution 15: 44–59.

van der Plank, J. E. 1960. Analysis of epidemics. Plant Pathol. 3: 229–289.

Vanderpool, S. S., W. J. Elisens, and J. R. Estes. 1991. Pattern, tempo, and mode of evolutionary and biogeographic divergence in *Oxystylis* and *Wislizenia* (Capparaceae). Amer. J. Bot. 78: 925–937.

van Dijk, H. 1984. Genetic variability in *Plantago* species in relation to their ecology. 2. Quantitative characters and allozyme loci in *P. major*. Theor. Appl. Genet. 68: 43–52.

van Dijk, H., and W. van Delden. 1981. Genetic variability in *Plantago* species in relation to their ecology. Theor. Appl. Genet. 60: 285–290.

Van Dijk, P. J., and T. Bakx-Schotman. 1997. Chloroplast DNA phylogeography and cytotype geography in autopolyploid *Plantago media*. Molec. Ecol. 6: 345–352.

Van Dijk, P. J., M. V. Hartog, and W. Van Delden. 1992. Single cytotype areas in autopolyploid *Plantago media* L. Biol. J. Linn. Soc. 46: 315–331.

Van Fleet, D. S. 1969. An analysis of the histochemistry and the function of anthocyanin. Advancing Frontiers of Plant Science 23: 65–89.

Van Houten, W. J. H., N. Scarlett, and K. Bachmann. 1993. Nuclear DNA markers of the Australian *Microseris scapigera* and its North American diploid relatives. Theor. Appl. Genet. 87: 498–505.

van Tienderen, P. H., I. Hammad, and F. C. Zwaal. 1996. Pleiotropic effects of flowering time genes in the annual crucifer *Arabidopsis thaliana* (Brassicaceae). Amer. J. Bot. 83: 169–174.

van Treuren, R., R. Bijlsma, N. J. Ouborg, and W. van Delden. 1993. The significance of genetic erosion in the process of extinction. III. Inbreeding depression and heterosis effects caused by selfing and outcrossing in *Scabiosa columbaria*. Evolution 47: 1669–1680.

Van Valen, L. 1976. Ecological species, multispecies, and oaks. Taxon 25: 233–239.

Vasek, F. C. 1958. The relationship of *Clarkia exilis* to *Clarkia unguiculata*. Amer. J. Bot. 45: 150–162.

Vasek, F. C. 1964. The evolution of *Clarkia unguiculata* derivatives to relatively xeric environments. Evolution 18: 26–42.

Vasek, F. C. 1968. The relationships of two ecologically marginal, sympatric *Clarkia* populations. Amer. Natur. 102: 25–40.

Vasek, F. C. 1977. Phenotypic variation and adaptation in *Clarkia* section Phaeostoma. Syst. Bot. 2: 251–279.

Veilleux, R. E., and F. I. Laver. 1981. Variation for 2n pollen production in clones of *Solanum phureja* Juz. and Buk. Theor. Appl. Genet. 59: 95–100.

Vekemans, X., and C. Lefebvre. 1997. On the evolution of heavy-metal tolerant populations in *Armeria maritima*: evidence from allozyme variation and reproductive barriers. J. Evol. Biol. 10: 175–191.

Via, S., and R. Lande. 1985. Genotype-environment interaction and the evolution of phenotypic plasticity. Evolution 39: 505–522.

Vickery, R. K., Jr. 1978. Case studies in the evolution of species complexes in *Mimulus*. Evol. Biol. 11: 405–507.

Vrba, E. S. 1980. Evolution, species and fossils: how does life evolve. South Afr. J. Sci. 76: 61–84.

Vrba, E. S. 1987. Ecology in relation to speciation rates: some case histories of Miocene-recent mammal clades. Evol. Ecol. 1: 283–300.

Vrba, E. S. 1989. Levels of selection and sorting with special reference to the species level. Oxford Surveys in Evolutionary Biology 6: 111–168.

Vrba, E. S. 1995. Species as habitat-specific complex systems. Pages 3–44 in D. M. Lambert and H. G. Spencer, eds., Speciation and the Recognition Concept. Johns Hopkins University Press, Baltimore.

Wade, M. J. 1996. Adaptation in subdivided populations. Pages 381–405 in M. R. Rose, and G. V. Lauder, eds., Adaptation. Academic Press, New York.

Wade, M. J., M. L. McKnight, and H. B. Shaffer. 1994. The effects of kin-structured migration on nuclear and cytoplasmic genetic diversity. Evolution 48: 1114–1120.

Wagner, W. L., and V. A. Funk, eds. 1995. Hawaiian Biogeography—Evolution on a Hot Spot Archipelago. Smithsonian Institution Press, Washington, D.C.

Wagner, W. L., S. G. Weller, and A. K. Sakai. 1995. Phylogeny and biogeography in *Schiedea* and *Alsinidendron* (Caryophyllaceae). Pages 221–258 in W. L. Wagner and V. A. Fink, eds., Hawaiian Biogeography–Evolution in a Hot Spot Archipelago. Smithsonian Institution Press, Washington, D.C.

Waller, D. M., D. M. O'Malley, and S. C. Gawler. 1987. Genetic variation in the extreme endemic *Pedicularis furbishiae* (Scrophulariaceae). Conserv. Biol. 1: 335–340.

Walsh, J. B. 1982. Rate of accumulation of reproductive isolation by chromosome rearrangements. Amer. Natur. 120: 510–532.

Wang, H., E. D. McArthur, S. C. Sanderson, J. H. Graham, and D. C. Freeman. 1997. Narrow hybrid zone between two subspecies of big sagebrush (*Artemisia tridentata*): Asteraceae). IV. Reciprocal transplant experiments. Evolution 51: 95–102.

Warwick, S. I., J. F. Bain, R. Wheatcroft, and B. K. Thompson. 1989. Hybridization and introgression in *Carduus nutans* and *C. acanthoides* reexamined. Syst. Bot. 14: 476–494.

Warwick, S. I., and L. D. Black. 1986. Genecological variation in recently established populations of *Abutilon theophrasti* (velvetleaf). Canad. J. Bot. 64: 1632–1643.

Warwick, S. I., and P. B. Marriage. 1982a. Geographical variation in populations of *Chenopodium album* resistant and susceptible to atrazine. I. Between- and within-population variation in growth and response to atrazine. Canad. J. Bot. 60: 483–493.

Warwick, S. I., and P. B. Marriage. 1982b. Geographical variation in populations of *Chenopodium album* resistant and susceptible to atrazine. II. Photoperiod and reciprocal transplant studies. Canad. J. Bot. 60: 494–504.

Waser, N. M. 1983. Competition for pollination and floral character differences among sympatric species: a review of the evidence. Pages 277–293 in C. E. Jones and R. J.

Little, eds., Handbook of Experimental Pollination Ecology. Van Nostrand Reinhold, New York.

Waser, N. M., L. Chittka, M. V. Price, N. M. Williams, and J. Ollerton. 1996. Generalization in pollination systems, and why it matters. Ecology 77: 1043–1060.

Waser, N. M., and M. V. Price. 1994. Crossing distance effects in *Delphinium nelsonii*: outbreeding and inbreeding depression in progeny fitness. Evolution 48: 842–852.

Washitani, I. 1996. Predicted genetic consequences of strong fertility selection due to pollinator loss in an isolated population of *Primula sieboldii*. Conserv. Biol. 10: 59–64.

Washitani, I., R. Osawa, H. Namai, and M. Niwa. 1994. Patterns of female fertility in heterostylous *Primula sieboldii* under severe pollinator limitation. J. Ecol. 82: 571–579.

Weaver, S. E., V. A. Dirks, and S. I. Warwick. 1985. Variation and climatic adaptation in northern populations of *Datura stramonium*. Canad. J. Bot. 63: 1303–1308.

Weaver, S. E., S. I. Warwick, and B. Thompson. 1982. Comparative growth and atrazine response of resistant and susceptible populations of *Amaranthus* species from southern Ontario. J. Appl. Ecol. 19: 611–620.

Webb, T. III. 1987. The appearance and disappearance of major vegetational assemblages: long-term vegetational dynamics in eastern North America. Vegetatio 69: 177–187.

Weber, E. 1997. Morphological variation of the introduced perennial *Solidago canadensis* L. *sensu lato* in Europe. Bot. J. Linn. Soc. 123: 197–210.

Weber, E. 1998. The dynamics of plant invasions: a case study of three exotic goldenrod species (*Solidago* L.) in Europe. J. Biogeogr. 25: 147–154.

Weber, W., and B. Schmid. 1998. Latitudinal population differentiation in two species of *Solidago* (Asteraceae) introduced into Europe. Amer. J. Bot. 85: 1110–1121.

Weller, S. G., A. K. Sakai, and C. Straub. 1996. Allozyme diversity and genetic identity in *Schiedea* and *Alsinidendron* (Caryophyllaceae: Alsinoideae) in the Hawaiian Islands. Evolution 50: 23–34.

Weller, S. G., A. K. Sakai, A. E. Rankin, A. Golonka, B. Kutcher, and K. E. Ashby. 1998. Dioecy and the evolution of pollination systems in *Schiedea* and *Alsinidendron* (Caryophyllaceae: Alsinoideae) in the Hawaiian Islands. Amer. J. Bot. 85: 1377–1388.

Wheeler, N. C., and R. P. Guries. 1982. Population structure, genetic diversity, and morphological variation in *Pinus contorta* Dougl. Can. J. Forest Res. 12: 595–606.

Wherry, E. T. 1955. The Genus *Phlox*. Morris Arboretum Monogr. III. Philadelphia.

Whitham, T. G., G. D. Martensen, K. D. Floate, H. S. Dungey, B. M. Potts, and P. Keim. 1999. Plant hybrid zones affect biodiversity: tools for a genetic-based understanding of community structure. Ecology 80: 416–428.

Whitkus, R. 1998. Genetics of adaptive radiation in Hawaiian and Cook Islands species of *Tetramolopium* (Asteraceae). II. Genetic linkage map and its implications for interspecific crossing barriers. Genetics 150: 1209–1216.

Whitlock, M. C., and D. E. McCauley. 1990. Some population genetic consequences of colony formation and extinction: genetic correlations within founding groups. Evolution 44: 1717–1724.

Whitlock, M. C. 1997. Founder effects and peak shifts without genetic drift: adaptive peak shifts occur easily when environments fluctuate slightly. Evolution 51: 1044–1048.

Whittemore, A. T., and B. A. Schaal. 1991. Interspecific gene flow in sympatric oaks. Proc. Natl. Acad. Sci. USA. 88: 2540–2544.

Widen, B. 1993. Demographic and genetic effects on reproduction as related to population size in a rare, perennial herb, *Senecio integrifolius* (Asteraceae). Biol. J. Linn. Soc. 50: 179–195.

Wiley, E. O. 1981. Phylogenetics: The Theory and Practice of Phylogenetic Systematics. John Wiley & Sons, New York.

Wilkinson, D. M. 1997. Plant colonization: are wind dispersed seeds really dispersed by birds at larger spatial scales? J. Biogeogr. 24: 61–65.

Williams, E. G., R. B. Knox, and J. L. Rouse. 1982. Pollination subsystems distinguished by pollen tube arrest after incompatible interspecific crosses in *Rhododendron* (Ericaceae). J. Cell Sci. 53: 255–277.

Williams, W. 1960. The effect of selection on the manifold expression of the "suppressed lateral" gene in the tomato. Heredity 14: 285–296.

Williamson, M. 1996. Biological Invasions. Chapman & Hall, London.

Williamson, M., and K. C. Brown. 1986. The analysis and modeling of British invasions. Phil. Trans. Royal Soc. Lond. B. 314: 505–522.

Wilson, P., and J. D. Thomson. 1996. How do flowers diverge? Pages 88–111 in D. G. Lloyd and S. C. H. Barrett, eds., Floral Biology. Chapman & Hall, New York.

Willson, M. F. 1993. Dispersal modes, seed shadows, and colonization patterns. Vegetatio 107/108: 261–280.

Wissel, C., and S. Stocker. 1991. Extinction of populations by random influences. Theor. Pop. Biol. 39: 315–328.

Wolf, P. G., D. E. Soltis, and P. S. Soltis. 1990. Chloroplast-DNA and electrophoretic variation in diploid and autotetraploid *Heuchera grossulariifolia*. Amer. J. Bot. 77: 230–242.

Wolf, P. G., and P. S. Soltis. 1992. Estimates of gene flow among populations, geographic races, and species in the *Ipomopsis aggregata* complex. Genetics 130: 639–647.

Woodson, R. E., Jr. 1947. Notes on the "historical factor" in plant geography. Contr. Gray Herbarium 165: 12–25.

World Conservation Monitoring Centre. 1992. Global Biodiversity: Status of Earth's Living Resources. Chapman & Hall, London.

Wright, S. 1931. Evolution in Mendelian populations. Genetics 16: 97–159.

Wright, S. 1938. Size of population and breeding structure in relation to evolution. Science 87: 430–431.

Wright, S. 1939. The distribution of self-sterility alleles in populations. Genetics 24: 538–552.

Wright, S. 1940. The breeding structure of populations in relation to speciation. Amer. Natur. 74: 232–248.

Wright, S. 1941. On the probability of fixation of reciprocal translocations. Amer. Natur. 75: 513–522.

Wright, S. 1951. The genetical structure of populations. Ann. Eugenics 15: 323–354.

Wright, S. 1969. Evolution and Genetics of Populations. Vol. 2. University Chicago Press, Chicago.

Wright, S. 1977. Evolution and Genetics of Populations. Vol. 3. University Chicago Press, Chicago.

Wright, S. 1980. Genic and organismic selection. Evolution 34: 825–843.

Wright, S. 1982. Character change, speciation and higher taxa. Evolution 36: 427–443.

Wu, L., A. D. Bradshaw, and D. A. Thurman. 1975. The potential for evolution of heavy metal tolerance in plants. III. The rapid evolution of copper tolerance in *Agrostis stolonifera*. Heredity 34: 165–178.

Wyatt, R. 1988. Phylogenetic aspects of the evolution of self-pollination. Pages 109–131 in L. D. Gottlieb and S. K. Jain, eds., Evolutionary Biology. Chapman & Hall, New York.

Yang, T. W., and C. H. Lowe. 1968. Chromosome variation in ecotypes of *Larrea divaricata* in the North American desert. Madroño 19: 161–164.

Yeh, F. C., and C. Layton. 1979. The organization of genetic variability in central and marginal populations of Lodgepole pine *Pinus contorta* ssp. *latifolia*. Canad. J. Genet. Cytol. 21: 487–503.

Young, A., T. Boyle, and T. Brown. 1996. The population genetic consequences of habitat fragmentation for plants. Trends Ecol. Evol. 11: 413–418.

Zabinski, C. 1992. Isozyme variation in eastern hemlock. Canad. J. Forest Res. 22: 1838–1842.

Index